STREAMING MEDIA DONE RIGHT

The Complete Guide to Media Servers and Their Connections

Philip Rastocny

Streaming Media Done Right

By Philip Rastocny

Amazon edition

Publisher: Grasslands Publishing House, 8526 Central Avenue, Brooksville, Florida 34613

ISBN: 0-9854081-4-6

First edition — June, 2017

Library of Congress Control Number: 2017909373

Discover other titles by Philip Rastocny at http://www.Amazon.com

Cover design by Philip Rastocny

Cover images provided by www.PresenterMedia.com

Images created by Philip Rastocny or public domain

Edited by Althea Rose

Trademarks

- **Amazon**™ and **Amazon Fire**™ are trademarks of Amazon, Inc.
- **AMD**® is a registered trademark of Advanced Micro Devices, Inc.
- **Android**™ is a trademark of Google Inc.
- **AirPlay**® and **Apple**® are registered trademarks of Apple, Inc.
- **BarraCuda**® is a registered trademark of Seagate Technology, LLC
- **Bluetooth**® is a registered trademark of the Bluetooth Special Interest Group
- **ChannelMaster**® is a registered trademark of Channel Master, LLC
- **Chromecast**™ is a trademark of Google Inc.
- **DLNA**® is a registered trademark of the Digital Living Network Alliance
- **FireCuda**™, **IronWolf**™, and **Skyhawk**™ are trademarks of Seagate Technology LLC
- **Google**™ is a trademark of Google Inc.
- **HDMI**® is a registered trademark of HDMI Licensing LLC
- **intel**® is a registered trademark of the Intel Corporation
- **iTunes**® is a registered trademark of Apple, Inc.
- **Thunderbolt**™ is a trademark of Apple, Inc.
- **Linux**® is a registered trademark of Linus Torvalds
- **Microsoft**® is a registered trademark of the Microsoft Corporation
- **Nakamichi**® is a registered trademark of the Nakamichi Corporation Limited
- **Netflix**® is a registered trademark of Netflix, Inc.
- **PlayStation**® is a registered trademark of Sony Interactive Entertainment Inc.
- **ROKU**™ is a trademark of Roku Inc.
- **SONY**® is a registered trademark of the Sony Corporation
- **Sun Microsystems**® is a registered trademark of the Oracle Corporation
- **The Abso!ute Sound**® is a registered trademark of Absolute Multimedia, Inc.
- **TOSLINK**™ is a trademark of Kabushiki Kaisha Toshiba, doing business as Toshiba Corporation.
- **TiVo**® is a registered trademark of TiVo Inc.
- **WD**® is a registered trademark of the Western Digital Corporation

Philip Rastocny

For the person whose personal passion led me to the high-end

Rudy Bozak

Table of Contents

List of Illustrations

Forward

Since the introduction of digital music to the consumer market way back in 1982, an elite community better known as audiophiles have constantly complained about its quality and urged the industry to make this form of reproduction more accurate. Their complaint was that in comparison to analog turntable-and-record systems, digital audio was edgy and shrill. Decades later, these issues have finally been addressed.

Now the convenience of digital recordings (both audio and video) has taken on acceptance by this elite community but to gain this acceptance the size of the files grew exponentially. A 3-minute digital song recorded back in 1983 used about 60 Mb of disc space and today this same song can occupy well over 10 or even 100 times that size. So storage of these huge music files meant looking at a method far different from the convenience of a CD, DVD, or Blu-ray disc.

The reason for this is because the way music is recorded. Back in 1983, the industry had to adopt a standard so that recording and playback methods would not be confusing from the buyer's perspective (will this disc work in my player?). The standard recording/playback format chosen was what is called 16/44 (we will get into the definition of that later) and locked into place for about 10 years.

Over the decades since 16/44 came to be, new high-resolution (high-res) formats have arrived all potentially improving the resolution of the recorded music. These improvements increase the dynamic range, reduce the noise floor, and widen the bandwidth of the old 16/44 standard. Video wasn't standing still either and its technology has advanced from the original 4:3 format to the current 16x9 4K format with 8K and even 16K formats being considered. This is good news for those who have dreaded the sound and sights of these old standards but this is also bad news since there is an additional cost for this improvement (you never get anything for nothing). The amount of storage space required has grown almost beyond belief.

Multi-Terabyte media servers routinely find their way into home entertainment systems and sophisticated software allows users to access these huge high-res songs (and video) from the convenience of their tablets and cell phones, even remotely. Play lists and similar advantages of compressed music files (*MP3*) have been carried into this high-end realm making the transition not only painless but also highly attractive.

Along with purchased-and-stored songs came streaming music (and video) services which were a natural addition to high-res digital music. Such services permit you to listen to music/video through an internet connection just as one does by tuning in a radio station in your car. Some of these services are completely free and offer compressed music for zero costs. Others also allow you to add a small subscription fee to improve the fidelity. Still other services even offer streamed high-resolution audio and video in formats that rival the highest resolutions available and all through the convenience of the Internet.

So what does this mean to you as the consumer? You guessed it. You scratch your head at its complexity and run back to your MP3s and CDs for the sake of simplicity. However, once you *hear* and *see* the difference between a high-end system and these old standbys, you may not be able to shake the addiction – and yes like anything high-res can become an addiction.

This guide is intended to be a high-level overview to the mysterious world of streaming media demonstrating the numerous paths it can take; it is not possible to explain how to make every connection for every device for every scenario. It can simplify these complex issues and show you what it takes to add streaming digital music and video to your existing system. It also maps out how you can logically and economically evolve your current system into a truly high-end performer.

So with this said, hold onto your wallets and let's look at this evolutionary process as it applies to you. Since there are multiple audiences here, we will break the approach into an evolving path. We will start with the low-budget path (I want the most value for the lowest cost), progress through steps on upgrading this starting point (what should I upgrade first), and finish with the off-the-shelf high-end path (here's my credit card number now just sell me what I need).

Hopefully after reading this transition path you will benefit from these progressive steps knowing that there are alternatives to what you may already understand. In other words you will understand how streaming media is done right.

With this overture under your belt, roll up your sleeves and let's get started!

Introduction

I personally started out listening to music on the radio and then as a teenager I moved to vinyl records and a turntable. Live music was an irregularity in my life but I always enjoyed acoustic performances best. I couldn't wait to go to church on Sundays primarily to hear its huge pipe organ echoing through the rafters. What I heard at this stage in my ever-maturing appreciation of audio was the huge differences between these live performances and their recordings; things just sounded "off" through any speakers I heard.

Over the decades that passed, my musical maturity grew as did my investment in the components of my home entertainment system. Before I knew it, I had thousands of dollars tied up in speakers, turntables, phonograph cartridges, wires, and software (here vinyl records) and there seemed to be no end to these expenses. What bothered me that although the sound was more accurate (it sounded *better* with the addition of each new piece), music still did not sound the same as a live performance.

As my vinyl records wore out and pops, ticks, and hiss grew with each playing, the introduction of digital music with its promise of perfect recordings forever had a distinct appeal. The proof of these promises would be in its sound and that was still the most important thing to me. The convenience and quietness of a CD had one thing going for it, but back in the mid 1980s music coming from a CD player to me sounded much like fingernails on a chalkboard. So I waited for this great promise to evolve into its perfection and stayed with my vinyl despite its drawbacks. After all, it took 10 years for Nakamichi® to transform 8-track tapes into a respectable cassette tape recording format and I assumed digital music should be no different. As the saying goes, you must first learn to crawl before you can run, right?

Figure 1 - Pioneer H-R100 8-Track and Nakamichi 1000 Cassette

Well what I thought would take 10 years in reality took much, much longer and about 30 years before digital music stopped "crawling" and started to "run." These early improvements in digital audio mainly focused on the *playback* systems rather than the *recording* formats; fixing the problems at the destination rather than its source. One manufacturer offered oversampling, another offered more stable clocks, another more precision conversion, and yet another offered analog power supplies. While each tweak *improved* the playback sound, still none compared to a live performance much less sounding better than my trusty reference: vinyl records.

As an analogy, the Hubble Space Telescope mirror fiasco comes to mind. Once put into service, its initial images were terribly out of focus. After a quick evaluation, scientists discovered a spherical aberration (mirror figuring) issue and the solution was to add something like "glasses" to quickly correct this problem instead of changing the mirror (that is, band-aiding the problem at the destination rather than fixing it correctly at its source, here a new mirror).

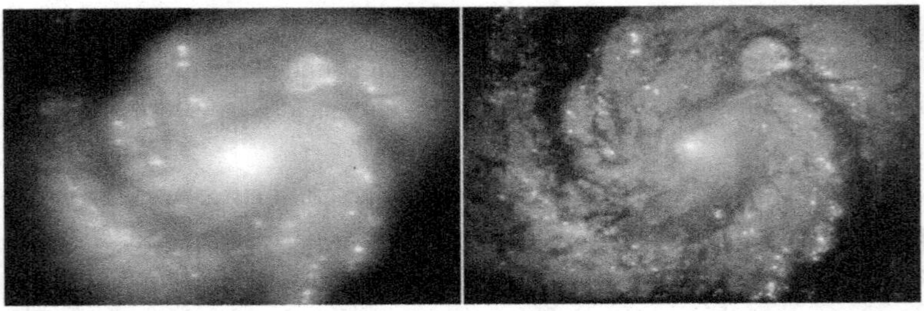

Figure 2 - Hubble Images Before and After "Glasses"

The next step in the digital music evolution came when the source recording formats changed. Digital music recordings are not inherently "wrong" as much as the way they were done (the format chosen to make the recording). Now instead of being locked into an archaic source format that was concerned about fitting over an hour of music on one disc, engineers created new recording methods that had no constraints as to the resultant file size. But this meant new studios had to be used to maintain this higher quality from end-to-end.

For example, if a studio moved to a better recorder, all components in the studio had to be changed to match (the "weakest link in the chain" scenario). Improving one link in this recording chain did little to improve the sound of a recording; the final quality is limited by the lowest-quality piece. In other words, a high-resolution recording passed through a low-resolution mixer yielded a low-resolution song. So more time was needed for all of this old – and very expensive – equipment to be upgraded (this is why it took over three decades for digital music to evolve).

Figure 3 - All Analog Studio Equipment Had to be Changed to Digital

Fortunately today, high-quality recordings can be made on an artist's own computer and a professional recording studio is no longer required to make really good sounding music. This is a boon to consumers like you who want to hear more than just the old canned and compressed releases created more out of habit than art.

Figure 4 - An Apogee Mini Recording Studio

Just as it is in a recording studio where moving from analog to digital was pretty expensive, moving from low-resolution to high-resolution digital is also not cheap. By definition, high resolution implies that the system playing a high-res source is also of very high quality (again the weakest link in the chain scenario). Fortunately, swapping out one piece of old gear is possible and a migration path can easily be planned out. In other words, you can choose which piece of gear to swap out first and get the optimum sonic improvement from that single change.

While playing CD source material through a high-resolution system can make these tracks sound *better*, for the best possible sound you must have an equally-good source material to feed your system (sort of like fixing the Hubble mirror instead of adding glasses). Services like Band Camp, Bravura, Burning Shed, and many others help young upcoming artists market their music and this keeps the cost of quality within reach of the average consumer while still helping to feed these artists. Other independent record companies like 2L, Blue Coast, Discogs, High-Res Audio, and a host of others all offer high-resolution downloads in various formats.

We will discuss many things along the way but the overall process when moving from one playback format to another does require *a lot* of change. Again, the first change you must embrace is the *massive* amount of disc space you will need to store and play these files and with it comes the loss of convenience and familiarity. If you are used to popping in a CD and listening to your favorite artist, this will become a thing of the past. The CD medium is woefully inadequate for storing these large high-resolution songs so alternatives to this medium must be used. As the saying goes, you cannot stuff 10 pounds of potatoes into a five-pound sack.

We will discuss alternative media sources, storage devices, and players that can help fill in the gap from what you now enjoy to what you will need in the future. We will also discuss the risk involved with the loss of your media (i.e., backup and reliability issues).

Along this line I want to mention that illegally copying music and/or video is just *wrong* no matter how you try to justify it in your mind. It's **stealing** period and although it is really easy to do in digital format, just don't do it. Every time you copy a song from a friend those artists who made that song starve a little more. Think twice before pressing COPY.

With this I believe we can get into the components of how this stuff works and point you in a sane direction of upgrading your system. We will begin from the start as if this were your first time into digital streaming and build on each topic.

As always, I encourage you to take your time, do good work, and have fun.

Philip Rastocny

Digital Recordings

Sound we hear in nature through our ears has always been analog. When you listen to a bird chirping in a tree, that's an analog sound. When you hear thunder in a rain storm rattle the glasses on your table, these are also analog sounds. Talking to your friends, honking your car horn, everything acoustic in life is an analog sound. The challenge arises when you try to take these analog sounds and preserve them in a digital format that recreates them accurately (an inaudible approximation).

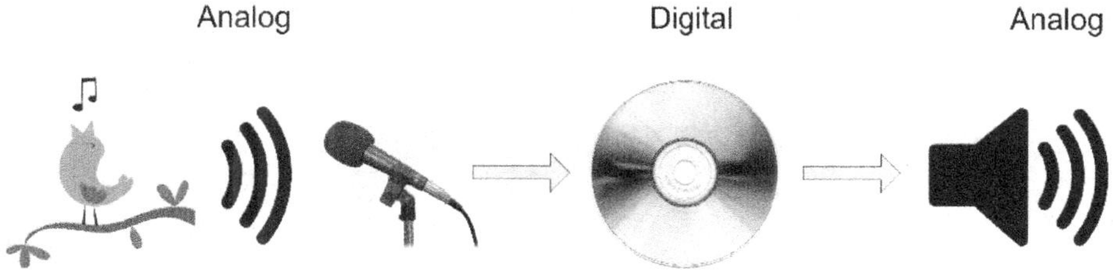

Figure 5 - Analog-to-Digital (A-D) and Digital-to-Analog (D-A) Conversions

The first methods used in digital recordings leveraged off of a computer's very high speeds to look at things in very small slices of time. The more slices (samples) you take, the more accurate the signal you can reproduce. But logic tells us that there is a point at which the number of samples exceeds the ability of the human ear to detect any improvement in quality. But how could you know what this rate was?

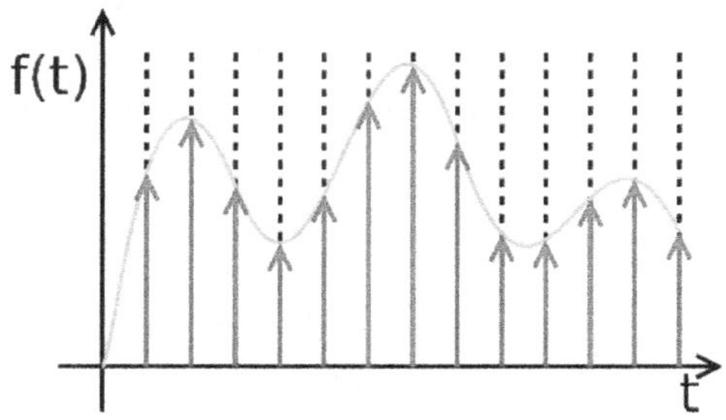

Figure 6 - Samples of an Analog Signal over Time

The scientific theory used to figure out how many samples to use – called the Nyquist theorem – was simple: use the upper limit of human hearing as the highest frequency you wish to preserve. That frequency used in the audio industry for decades was 20,000 cycles per second (20 Kilo Hertz or 20 kHz). Nyquist proved mathematically that when you sample at twice that rate (i.e., 40 kHz), you can reconstruct the analog signal without losing any information. Engineers added 10% to this assertion and so a standard of 44.1 kHz was born. Without passing judgment on this standard (there is *a lot* of dissention on this standard), let's continue.

Next came the question of how loud of a signal should be preserved (the difference between no signal and the loudest signal). This is called the signal's dynamic range and for digital music it is called its *bit depth.* The best analog recordings of the time had a dynamic range limited by the movement of a cutting head on machines used to create vinyl records (about 70 dB). So digital audio had to have a dynamic range of at least of 70 dB (of course more would be better).

Dynamic Range is Like the Difference Between a Whisper and a Roar

Figure 7 - Dynamic Range Makes the Difference

The limiting factor in this decision was the fixed amount of data that could be stored on the medium (the capacity of a standard audio disc or 640 Mb, a size chosen to fit into a single-DIN standard slot in an automobile). Here you have two choices, where one fixes the other: 1) the maximum recording time, or 2) the size of the sampling rate (sample size or bit depth). A maximum recording time of 74 minutes was chosen and that fixed the size of the sample to 16 bits translating to a dynamic range of 90 dB. For a complete description involving a lot more than this high-level explanation, the CD-DA description on Wikipedia or the complete story behind the CD-DA in the Journal of the Audio Engineering Society will alleviate your curiosity.

Everyone was pleased, engineers and salespeople alike, and Sony/Philips published this specification in a booklet with a red cover (this is why this standard also known as a Red Book CD). Ever since, all CD-DA pressings are recorded using a 16-bit depth with a 44.1 kHz sampling rate (abbreviated as 16/44).

Figure 8 - 1979 Announcement of the CD-DA

This locked in a technology that took off faster than even the most optimistic marketing estimations. Its convenience, durability, and silence (lack of pops and ticks) proved more appealing than the audio quality it provided. And since CD-DA recordings could hold more music and whose discs could be used in a car stereo, it created an instant market in both the home and automotive industries. It's no wonder that this standard took off.

The high-end audio industry, however, was unreceptive to this new format. Complaints arose regarding edginess and shrieking not to mention an unnatural sound so it took a while for design engineers to first believe that there was an actual problem and then second to figure out why digital music sounded good on paper (in theory) but not so much in reality.

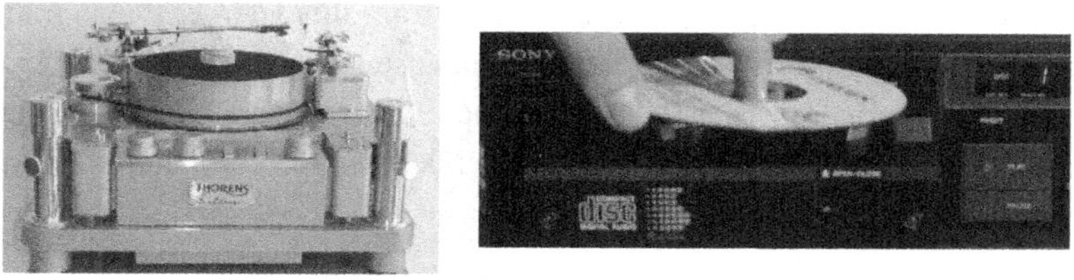

Figure 9 - High-End Critics Preferred Analog Sound over Digital Sound

At first, nothing was done to this technology where skeptics to this criticism cited data and critics cited subjective sound. This is essentially a stalemate since these two camps speak different languages and neither understood the other. Much like a right-brained and left-brained war, it appeared that nothing would ever change, but when the phase of the audio signal was studied and proved severe, it did. Efforts to improve digital sound were initially focused on what is called an anti-aliasing filter, the thing that keeps unwanted digital harmonics from impacting the quality of the decoded sound. But in reality there was a deeper understanding of the problem than just this filter design could explain.

Alternatives to 16/44

Another of the main complaints was the fact that digital music would go "deaf" meaning wide-dynamic-range recordings suddenly just become silent. Recordings of live music easily showed this failing where dying echoes in concert halls and organ reverberations in churches would at a point just abruptly disappear. This was a miscalculation of the dynamic range required to make an accurate recording so the size of the sample (bit depth) needed to be increased. Here the first number of "16" was soon replaced by 20 and eventually 24.

But as previously mentioned anti-aliasing filter had issues of its own and the only way to truly resolve this issue was to increase the sampling rate and thereby putting this frequency at multiples of the original 44.1 kHz rate (88.2, 196.4, 384.8). Higher sampling rates yielded less-complicated filter designs and the sound became instantly more appealing. As a result, several formats were created and the most common today is 24/96, a good format that resolves most audiophile's complaints about the so-called "digital sound."

During this 30-year period the analog industry was not sitting on its laurels and it also studied ways to improve the sound from this source. Phonograph cartridges, turntables, and tonearms all went under intense scrutiny and radical designs resulted with astronomical price tags. Some people still prefer analog over high-resolution digital recordings claiming all-analog versions sound the purest and most accurate when compared to the real thing.

Encoding/Decoding Formats

With digital audio, there are literally an infinite number of ways you can store and play back music. All you have to do is create a new standard for encoding the music while recording and then another standard for decoding that same music on playback. These are called CODECs (short for coders-decoders). Fortunately, creating a CODEC is a complicated issue so their numbers are not as numerous nor as widely accepted as one may think. (See *Abbreviations and Glossary* for a partial listing of available digital audio formats.)

One feature in digital technologies commonly used to save space is to compress the data (such as a computer's ZIP files). CD-DA recordings are not compressed so finding a way to make these big files smaller soon found its way into the audio community. Although inherent losses to audio quality are always encountered whenever compressing the original music, what you gain with compression is more music in the same space. In effect, you can put your entire music library on hand-held devices such as cellular phones and dedicated players. One can go jogging with a player half the size of a business card clipped to your shirt and listen to your own music library for hours at a time never hearing the same song twice.

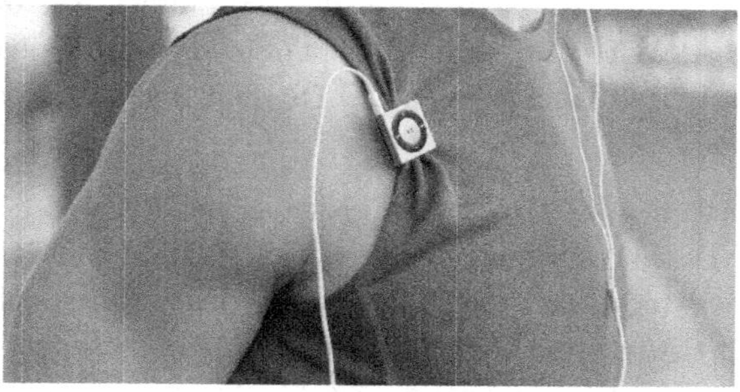

Figure 10 - Compressed Music Files Are Very Handy

Compressed music was also quite saleable so download services thrived offering the latest albums instantly; the Apple® iTunes® service probably being the most familiar. Apple also decided to add software to their players allowing you to create *play lists* that selected certain songs from certain albums in whatever order you desired thereby allowing you to create a flow of music for a certain preference or mood. This idea took off. Other companies followed suite and another industry was born to feed the ever growing demand for what was termed *streaming music*.

The most common music compression standard used today is the MP3 format but many others exist. For example, Apple uses its own standard called Advanced Audio Coding (AAC) and is available on their iTunes download site. But other uncompressed formats also exist including another lossless format created by Apple known as ALAC (Apple Lossless Audio Codec). Other formats grew such as Microsoft® WAV, Sun Microsystems® AU, and another yet another Apple format called AIFF.

As mentioned earlier, there is nothing to keep someone from creating a new recording or playback format; you just need to build an encoder and decoder (CODEC). An independent effort by a group of folks not interested in making money on their design created an open-source format call the Free Lossless Audio Codec (FLAC) that is highly popular in the high-end community. As it is when ripping a CD to a compressed MP3, this CODEC rips a CD to an uncompressed FLAC. But FLAC is still much like putting on glasses rather than fixing the mirror since the preserved quality is only 16/44.

Still evolving is yet another uncompressed format called Direct-Stream Digital (DSD). Here the approach to recording is completely changed. DSD recordings use a single-bit depth and a sampling frequency of 2.8224 MHz. This technique puts the design of the anti-aliasing filter in a different league from other formats and reportedly improves the sonic accuracy again. DSD format was originally used in Super-Audio CDs (SACDs), an earlier attempt to provide better quality music using the same physical sized medium as a CD-DA, but today is available from may download retailers.

As mentioned, the DSD format is evolving and today it is available in its original version (DSD64) or higher-end versions (DSD128, DSD256, and DSD512). The suffix numbers for a DSD song tells you how many times higher the sampling rate is over a CD-DA recording (64, 128, 256, and 512 times higher) with sampling rates of 2.8 MHz, 5.6 MHz, 11.2 MHz, and 22.4 MHz respectively. The upside of DSD files are their acclaimed improvement audio accuracy and the downside is that file size for each song is staggeringly huge. With the higher sampling rates, the ultrasonic noise associated with 16/44 recordings is minimized in DSD recordings. But with the price of multi-terabyte disc drives dropping daily, DSD recordings are becoming more and more popular.

I love standards because there are so many from which to choose…

The downside of trying to improve audio recording quality over the original CD-DA standard is that all of these efforts in creating better sound are incompatible with each other. For example, DSD64 players cannot decode DSD128 music but DSD128 players can decode both DSD64 and DSD128, etc. (downward but not upward compatible). Fortunately, most players available today include multiple decoder capabilities and today music encoded in high standards can be decoded on even modestly-priced portable units.

Figure 11 - Neil Young's PONO Music Players Decode High-Res FLAC

To make a very long story short, you must decide on an audio format so you can decide how to build your system to support that format. From the current trend there is one thing you can count on: music file sizes are getting very, very large. What once took up 640 Mb of space for an entire CD-DA album now requires over 2 Gb in DSD64 and multiples of this for higher DSD rates.

What Format Should I Choose?

Regardless of the recording format wars to which we are currently subjected, the fact remains that digital world is here to stay. The point of this book is to explain how to get the best sound possible regardless of the format you choose. There is much to consider when venturing into steaming audio/video realm the first of which is to decide on your favorite format. But remember that once you decide, you can make a low-resolution format sound better but it will never be as good as a high-resolution source file.

Personally, I prefer uncompressed formats such as FLAC and WAV but the logical side of my brain tells me to look at DSD as the high-end format of the future and plan ahead accordingly. As it was with the introduction of CDs, I bought 16/44 versions and then later bought 24/96 versions. Currently, I only buy DSD64 or higher versions if they are available. The truth of the matter is this: as time moves on and a better understanding of how to more faithfully capture and reproduce analog signals to/from a digital format, things will change and there is no telling what format in the future will sound the absolute best (the most like the real thing). For this reason, I encourage you to be flexible; plan for change and know that whatever you buy today may be replaced in the near or distant future.

I do have one piece of advice: if I were you, I would no longer buy *any* compressed music since the cost of storage space today is not an issue and a compressed music file by definition contains many losses. And even if you still want to listen to compressed music in your car or your portable device, free software allows you to convert file formats from one to another (such as Foobar2000). This allows you to keep your original purchase in high-resolution format to enjoy on your high-end audio system while still offering a lower-resolution format for less-demanding listening environments. Such an approach just takes up a little more disc space and a little time to make this conversion.

For consistency's sake, I will discuss DSD audio (and MKV video) as the current source formats you wish to preserve. The only thing you need to remember is that if you choose a format with an inherently larger/smaller file size, all you need to do is scale up/down the size of the storage space to match.

With this in mind, let's first talk about the components required to play back digital audio files.

Playback Components

Playing digital music is *far* more complicated today than it used to be back in the 1990s. To play a CD-DA recording back then, all you needed was a CD player for your home and another for your car. You put a disc in the tray, pressed play, and out came music. Wow! What a welcomed change from those noisy huge vinyl discs that had to be treated with kid gloves or those cassette tapes that broke and jammed inside their players.

Figure 12 - The Challenges before Digital Music

As time progressed, these digital players could also decode compressed music formats (e.g., MP3) allowing you to burn your own discs with whatever music you desired and listen to dozens of albums on one disc. Soon thereafter hand-held MP3 players made their way into the market with solid-state drives. These allowed you to listen to music while you jogged or mowed the lawn. To rip a CD to MP3 and copy it to such a device was also a no-brainer. Now that was incredibly convenient!

Figure 13 - An Apple iPod from 2001

Unfortunately, these evolutionary processes focused more on the convenience and less on the quality of these devices so their fidelity was nothing to get excited about. But as mentioned earlier, music recording and playback formats were also evolving at this same time. And in the high-end realm, the components used to playback these traditional and evolving formats broke out into individual pieces.

At their introduction in the 1980s, a CD-DA player included everything in one (*all-in-one*) box: a power supply, computer, control software, a drive bay, and a digital-to-analog decoder (DAC). One simple cable connected this player to a receiver. But when building high-end CD-DA players, engineers separated each of these pieces into their own chassis and cabling became more complicated. Understanding how each of these pieces fits together is where our journey begins.

The Basics

If all you want to do is stream music from MP3 files, most home receivers (and even smart televisions and satellite receivers) can do that for you now. Just connect a USB drive to your PC, copy your ripped files over to this drive, plug this drive to the input of your playback device, and viola: you are streaming digital music. This playback device is also known as an all-in-one device.

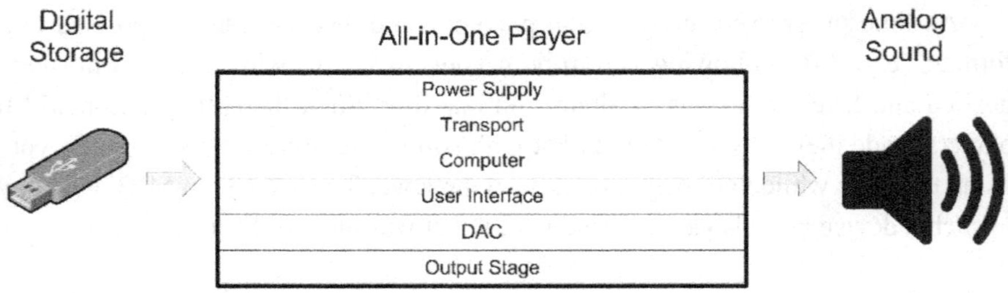

Figure 14 - Building Blocks of an All-in-one Device

Some all-in-one (Ai1) devices also offer wireless Bluetooth connectivity making your hand-held player its own "music server." But doing so puts you at the mercy of the quality of those individual pieces chosen in the design of that Ai1. Another issue is that the format you wish to play may not be supported by your Ai1. Saying this another way, there is no way to upgrade an Ai1 device; your only recourse is to buy a new Ai1 device that does.

You may already have your music library on an SD card/memory stick in your smart phone or loaded into the on-board memory of your hand-held music player. You may even stream your library from such an Ail device to your home entertainment system via a dock, USB port, or wirelessly.

Figure 15 - A Smart Phone is One Type of Ail Player

Let's assume that you have a wireless device on which you already listen to streaming music and that you use the Bluetooth connection to your system to listen to this music. What are the "behind the scenes" pieces required for this to happen? Much like changing tires on a car can change the comfort or performance of its ride, replacing these individual pieces – where possible – can improve the quality of that part and this translates into higher fidelity to your ears.

The Chain-Link Analogy

Maintaining a high-quality signal from the microphone in the recording studio to the speakers in your audio system is a very long journey through wires, transistors (and sometimes tubes), and who knows what else. The point is this: the quality of the audio signal you finally hear (or the video signal you see) is depreciated by every single thing encountered on its trip from the original microphone (including the microphone or camera). If one piece of equipment encountered on this trip is of inferior quality, the resulting sound or image is reduced to that level and cannot be restored. This is known as "the weakest link in the chain" analogy.

Figure 16 - Every Component in your System is Like a Link in a Chain

Putting it another way, **you have the opportunity to improve the quality of the things under your control**. Now that's a positive way to look at the issue and one that high-end audio and video is all about. But finding the weak link in your chain can be difficult and elusive. Some folks just start swapping things in and out to see if they can hear or see a difference but a down-stream weak link may hide or conceal an up-stream improvement. The most logical place to start is at the last link (your speakers or video display). But there is one more link between you and the source and that is your ears and eyes.

The very first thing you should do before starting to swap things out in your pursuit of finding the weak link in your chain is to have your ears professionally cleaned. Although you may not need this done, having them cleaned removes any doubt that there is something wrong with your hearing that contributes to your inability to detect up-stream changes. A professional irrigation takes time from your busy day but is well worth the effort if you have never had this procedure done. In the high-end audio world, attention to detail is always important so see our doctor and get this maintenance detail out of the way.

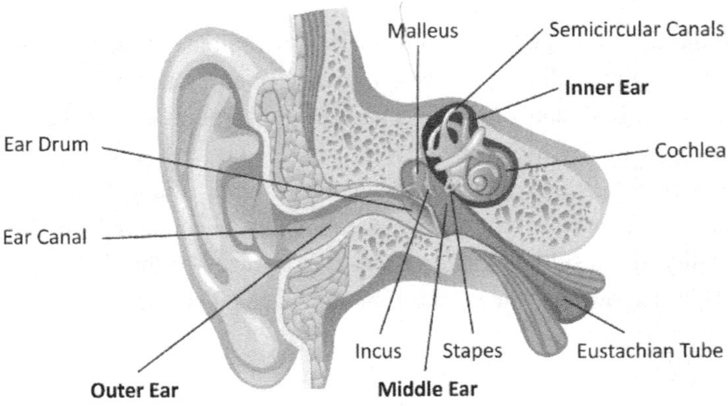

Figure 17 - Your Ears are the Last Link in the Audio Chain

So the process of eliminating sonic degradation issues should always begin backwards from your ears to the music source. The same logic should be applied to your eyes. Have them regularly checked professionally and if you need glasses, do yourself a favor and get some really good lenses that are color accurate. For now, let's focus on the audio chain.

Now that you are confident that you can hear properly, the other thing you may want to do is remind yourself what your favorite instrument sounds like live. After all, what you are really trying to do is to make your system sound real as compared to sounding good, right? My favorite instrument is the piano and its big brother the pipe organ. So to keep me sane, I just listen to one every now and then just to remind me how they really sound and I strongly suggest that you do the same. From this point of reference, you can more easily discern what your system reproduces and what it *should* reproduce.

Figure 18 - Know What Your Favorite Instrument Sounds Like

Without a doubt, the biggest change you can hear in your system is by changing out your speakers. As the first link in your audio chain between your ears and the music, the plethora of speaker designs available to audiophiles is vast. Each type of design offers something the other cannot. For example, if your ears are sensitive to attack (the beginning of a note) you may prefer the sound of a horn loudspeaker. On the other hand, if your ears are more sensitive to the human voice, you may prefer the sound of a planar speaker. Changing your speakers to the type of music you prefer will reveal more of what your ears are listening for.

Figure 19 - Planar, Horn, and Dynamic Loudspeakers

The next steps work up-stream from your speakers starting with speaker wires and their connections. While you can spend huge amounts of money in this area, moving up from thin wires to those of more substantial size will minimize the losses incurred in these thin wires. Without getting into the battle of what is and what is not audible, if you have not already done so, just keep an open mind and try the experiment of swapping out your speaker wires. Spend a few dollars and listen to what you already have in the rest of your audio chain to see if wires make things sound better. If so, keep what you've bought; if not, it was the cost incurred in the process of elimination. In this process you determined that the weak link was not (or no longer is) your speaker wires and you can move upstream to the next piece.

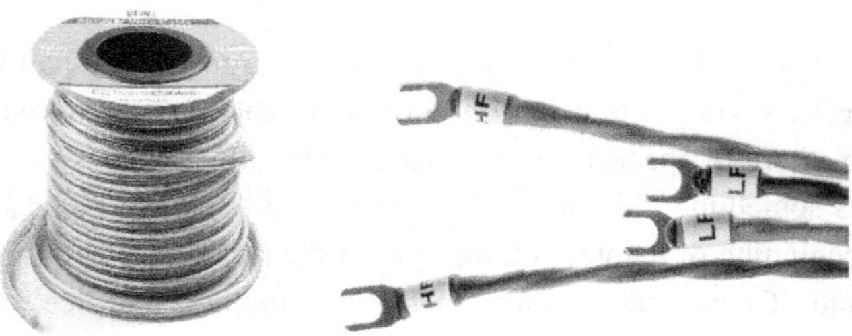

Figure 20 - Try Different Speaker Wires

Speaker wire connections are another area where weak links can be encountered. If your speaker wires are twisted around a terminal, this may "work" but a better way to ensure a positive connection is to use some sort of professional connector or lug. And the connection should be positive and hand tightened to the terminal as opposed to push-button or spring-loaded. Spade lugs on wires and twist terminals on speakers and receivers are my personal favorites since they provide the most contact area at the connection. Have a technician change out your terminals if you need this done.

Figure 21 - Use Good Speaker Terminals

You can perform these same simple steps of improving things upstream of your ears all the way back to the source for your current system. Changing things out one piece at a time gives you the best understanding of where your weak links are. But know this: there can be interaction between downstream and upstream components so this search is not a simple one. In other words, settling on speaker wires for your current receiver may or may not sound accurate when swapping out a new receiver.

Figure 22 - Swapping Interconnect Cables is a Simple Change

Take your time and as always compare what you hear with your favorite instrument. When you settle on getting your current system tweaked (hopefully an optimal tweak), then think about changing the streaming source.

So where do you go from here? As it is with fancy automobiles or watches, there are numerous upgrade paths from which to choose depending upon your budget and personal priorities. To some, the biggest bang for the buck is the highest priority while with others it is the best possible sound regardless of cost. Most folks fit somewhere in between and this is where a reasonable plan for future growth and sonic improvements comes into play (get your "ducks in a row"). A planned approach spreads the cost of your system improvements over time and allows the technology to advance while making huge strides in playback quality.

Figure 23 - Plan Your Upgrade Path

Where to begin a planned approach depends entirely on what you currently own and where your own priorities are. Since we must start somewhere, let's begin by analyzing those component parts (building blocks) of an existing Ai1 device to see what is possible. This "deconstruction" analysis can be very revealing in understanding where the next upstream weak link hides.

Building Blocks

As mentioned earlier, the USB-drive route is the simplest way to stream music and will get you into digital audio reproduction at the lowest possible cost. Instead of using an all-in-one (Ai1) device to do this, another approach is to use individual components for each of the building blocks (i.e., deconstruct the Ai1). This is a higher-priced but far more flexible path and may prove to be more cost effective in the long run.

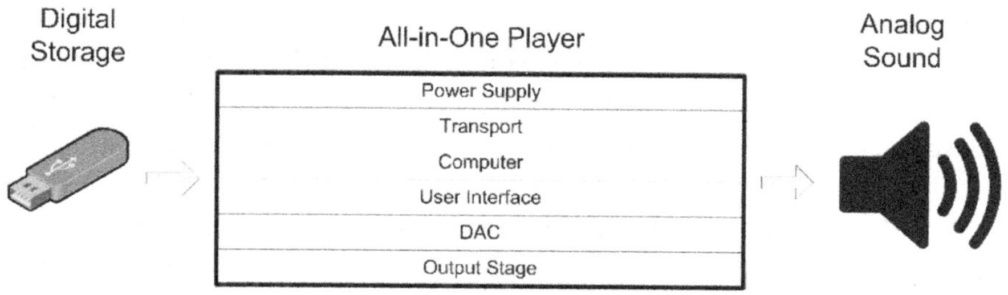

Figure 24 - Building Blocks of an All-in-One Device

If your existing Ai1 device cannot decode the format you have chosen to store your high-resolution audio files (e.g., plays FLAC but not DSD), you must make a break from the current low-resolution way you listen to music and embrace your high-resolution goal. This does not mean that you must toss your existing Ai1 device; just demote it to your secondary music source.

You may have an emotional attachment to your existing Ai1 device since it has served you well and you have mastered its features. This took time and most likely you must do this once again. Like buying a new car and getting used to its controls, features from your old car may not be available in your new car. The reason you re-learn something is because you want something else and sometimes you are willing to give up this to get that.

This also may be true when moving from your existing Ai1 device to your new one. What typically is involved is either relearning how to use it or giving up something to get something else. For example, your existing Ai1device may have certain features that you wish to retain, most likely your play lists. But when moving from a low-resolution format to high one, you may need to rebuild them.

Assumptions

In the process of deconstructing or the separation of the pieces of your Ai1 device, I will use DSD64 as the desired high-resolution format. Your choice may be different but the building blocks outlined herein remain the same.

There are many ways to deconstruct an Ai1 device for a relatively painless migration path. In other words, you can sometimes integrate an outboard component to your existing Ai1 device to improve its sonic quality without having to spend a lot of money (eliminate the next upstream weak link, this time one *inside* your Ai1 device). For a starting point, let's look at how these individual pieces fit together to better grasp what you may be able to change. Know that not all Ai1 devices are created equal and some cannot be enhanced by the methods suggested below.

It is also assumed that you currently own a home entertainment system on which you play CDs, tapes, or even stream music in some manner. This may be an A-V receiver, Smart TV, or a simple stereo with an AUX input (i.e., you listen to music through the air instead of ear buds). This next description explains where you may be able to begin your upward-migration path.

The Path

If it does not have the ability to handily manage your audio library, the first thing that must go is the Ai1; you will replace it with what is called a *music server*. Even if your Ai1 currently meets your needs, you will probably find yourself quickly migrating to a music server once you fully understand its benefits. There are many music servers on the planet from which to choose at many different price points but all do the same thing:

1. Allow you to access a storage device on which your music files reside (here a USB flash drive or external USB hard drive or even a USB connected Smart phone)
2. Decode the music format you have chosen (decode the digital DSD music files to a analog signal your receiver understands)
3. Stream this decoded music to your audio system (send this decoded analog signal out to your receiver)

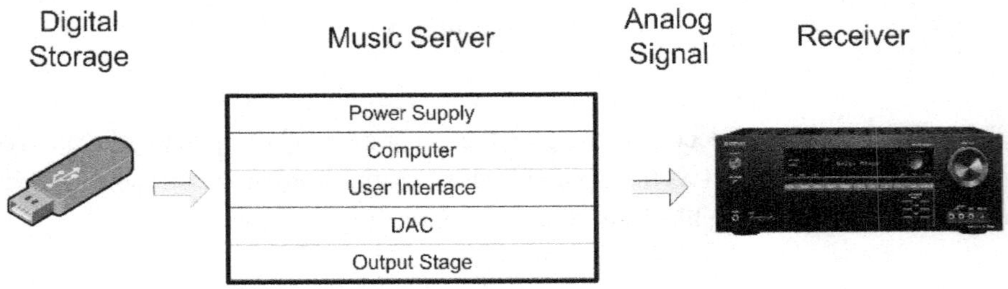

Figure 25 - Building Blocks of a Music Server

You can also take this same 3-point concept and apply it to streaming video. Here the only difference is that instead of an analog signal connected to a receiver, you use an HDMI® cable from what is termed a *media server* (*Home Theater PC* or *HTPC*). Just to keep things simple, we will explore music servers first.

So far we have been talking about simple USB connected digital storage devices better known as Disk-Access Storage or DAS to which almost any computer can be connected. USB ports come in two flavors: early styles of USB 2.0 (and older), and newer styles of USB 3.0 (and newer). Newer styles offer faster data transfer rates for the same connection than older styles but both work fine for audio applications. I still personally prefer USB 3.0 ports to more easily manage any parity error issues that may arise without affecting audio performance or delays (pauses or skipping).

If you have an old computer or laptop, you are at a pretty good starting point for migrating to a music server since you can use this old computer as the heart of the streaming audio/video system. You could also recruit your daily-use computer as the heart of your music server but moving files it in and out of your home entertainment system may become a bit of a hassle. If this is the case for you, I would plan on obtaining a dedicated computer for this purpose in the near future. What computer you buy is up to you but a laptop gives you everything you need in one small space (monitor, keyboard, and computer in one). All but the smallest will provide you with enough power from the USB ports to drive and control your connected devices.

What is required in this approach is the following:

- Computer or laptop (Operating System preference does not matter)[1]
- Interconnect cable from the sound card output to your receiver's external audio input
- (optional) Interconnect cable from your receiver's line-level (or TAPE) output to your computer's line-level (or microphone) input
- Control software (such as the JRiver Media Center or iTunes)
- Wireless network
- Tablet or cell phone

---

[1] This can depend upon which control software you decide to use

That's it! Without getting fancy and breaking things out more granularly, this is all you really need to get started. Again, you are at the mercy of the quality of the individual components you use (again the weakest link in the chain scenario), but this approach also allows you to swap things out and develop a migration plan for future improvements. Once you get started down this road, you may find yourself trying a friend's piece of equipment in your system (try before you buy) to understand where your next weakest link hides.

So how do you assemble these pieces? There are many ways to do so but again let's start at the simplest way first and look at the more complex ways down the road.

Step 1: Basic DAS Media Server

The most common type of music server available today uses either a built-in storage device or a storage device that is directly connected via a USB cable. Other direct connection types are Thunderbolt™ for Mac or eSATA and Firewire for Windows computers. These are all called *Direct-Access Storage* (DAS) devices since no additional components or complicated cabling arrangements are required between this type of storage device and a computer, however the external storage device may require a plug for its power supply.

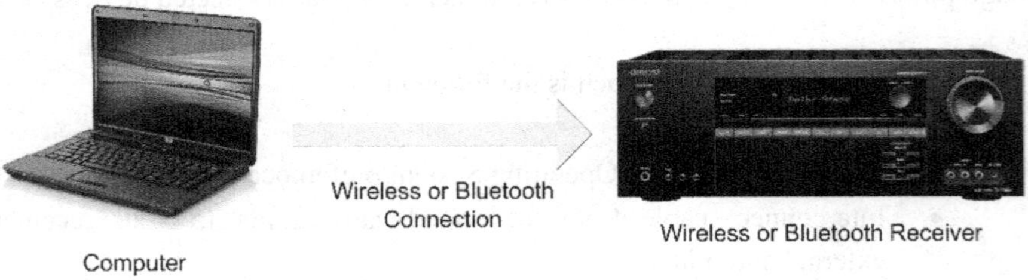

Figure 26 - A Wireless DAS Media Server

This first arrangement features a wireless or Bluetooth connection between the media server (computer) and the home entertainment system (receiver). Making a paired Bluetooth connection between the receiver and the computer is just like pairing any other two Bluetooth devices.

The media server software pretty much does all the rest. Similarly, a wireless connection streams whatever audio the computer supplies; optionally, these same wireless or Bluetooth connections allows your cell phone or tablet to select what music you wish to hear. The administration of software varies greatly from manufacturer to manufacturer and is beyond the scope of this document. But know that such software exists and there are many from which to choose. My personal preference for a music server application is the highly flexible and easily managed *JRiver Media Center*.

Step 2: Hard-Wired DAS Media Server

Taking this approach one step further adds a hard-wired connection between the computer and the receiver, something again that I also prefer over any kind of wireless connection. This involves using the output of the computer's built-in sound card and an unbalanced cable connection to the receiver.

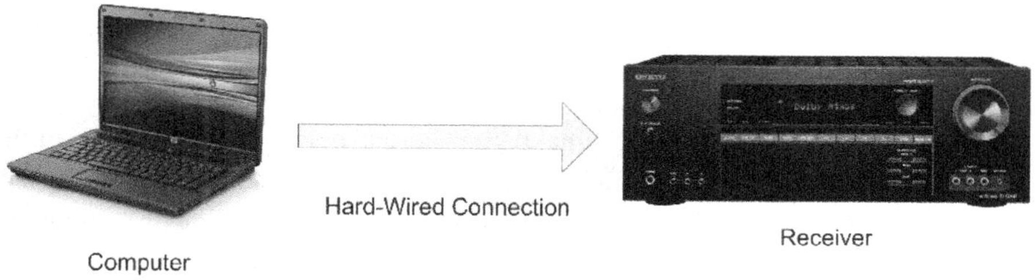

Figure 27 - Hard-Wired Media Server

The advantages of the hard-wired connection are in the cable's noise immunity and wider bandwidth. The downside is that the cable used must be long enough to reach between the computer and the receiver. If your computer is not normally located near your receiver this may not be an option at this time. Add to this issue of real-estate considerations, long high-quality interconnect cables are expensive and this route may be cost prohibitive at this time.

Figure 28 -A Good-Quality Hard-Wired Audio Cable

Step 3: External DAS Music Server

A basic DAS Music Server uses the built-in drive of the computer to store digital music files. Another option is to attach an outboard digital storage device via a USB, eSATA, FireWire, or Thunderbolt cable to the computer. This approach allows you to have more storage capacity than the built-in drive allows and can be easily removed in the event you wish to take your music library elsewhere.

Figure 29 - DAS Music Server with External Storage

I personally prefer using an external digital storage device simply because of the number of options available regarding its capacity and configuration. We will explore those options later but for now know that such an external device could also be a USB thumb drive.

With the Super-Speed USB 3.0 connections, data transfer rates to-from your PC and the USB drive are an excellent match not only for superior audio but also streaming video.

Figure 30 - Super Speed USB 3.0 Cable

If you have a lot of files and data scattered across hard drives, old computers, and other devices, you may want to use this opportunity to upgrade to a Network Storage Device (NAS). The great thing about a NAS is that you can put all of that data there instead and then *every* computer on your network can access it.

Step 4: NAS Music Server

The next step up in streaming music is to use a Network Access Storage or NAS device. A NAS device is remotely accessed through either a wireless or hard-wired Ethernet connection on your home network and can be physically located anywhere. There are serious upsides and complicated downsides to a NAS as we shall soon see.

For best operation NAS devices must be certified to a protocol called DLNA® (Digital Living Network Alliance), a sort of universal plug-and-play protocol for networked-attached devices. The upside of using certified computer operating systems, control applications, and peripheral devices certified to DLNA standards is that you will be able to seamlessly access and manage your library. The downside is that without end-to-end DLNA certification, it may still possible to listen to your digital music on your NAS; you must browse your library as if you were searching for a file on your computer.

Figure 31 - DLNA Makes Connectivity a Snap

Be aware that some NAS devices use a "compact" or "tiny" version of DLNA that are not certified or compatible with other certified devices. In these cases, you may not be able to access your music library with either your music server computer or your software application. Just remember to get a DLNA-certified device and all will be well.

As mentioned, one of the biggest features of using a NAS is that your music server computer can reside in one location and your NAS in another. Some DAS and NAS devices can be noisy or run hotter than you would prefer for your music room so locating them somewhere else may be the best solution. A DAS can be relocated only to a place that a USB cable can reach. A NAS can be located literally anywhere an Ethernet cable or reliable wireless connection can be reached.

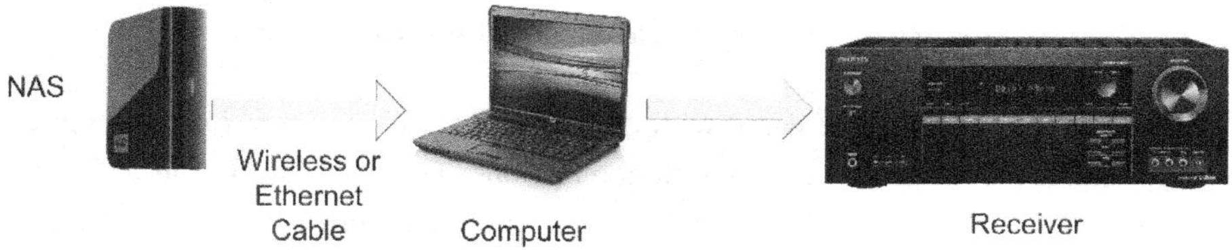

Figure 32 - NAS Music Server

I personally use hard-wired connections in my system. I have found that the background noise level is considerably lower when moving from wireless to hard-wired connections. The noise level in my listening room is quite low and I am able to detect these very minor changes; your listening room may be noisier and you may not notice an improvement moving from wireless to hard-wired connections.

Some Macintosh computers more often than not have issues maintaining reliable connections to NAS devices and therefore are most reliable when connected to DAS devices. If you are a Macintosh lover, you may wish to avoid a NAS since you may find yourself rebooting your computer a lot to restore its connection. Windows and Linux machines, on the other hand, are far more reliable in maintaining network connections and are good candidates for NAS devices.

The biggest downside to a NAS device is completely independent of the type of operating system used by the computer as we shall see next.

Hard-Wired Network Issues

Rarely in life will you get anything for nothing and with hard-wired network connections this is very true. Most home networks are historically wired with Category-5 (CAT-5) or CAT-5e Ethernet cabling as is most of the network-connected devices (routers, hubs, switches, modems, wireless access points, etc.). Noise immunity is the biggest issue with these types of networks and for this reason CAT-7 and CAT-8 networks are the quietest. Both of these network types require end-to-end cabling and end-to-end devices certified to this same standard to maintain the lowest possible noise floor.

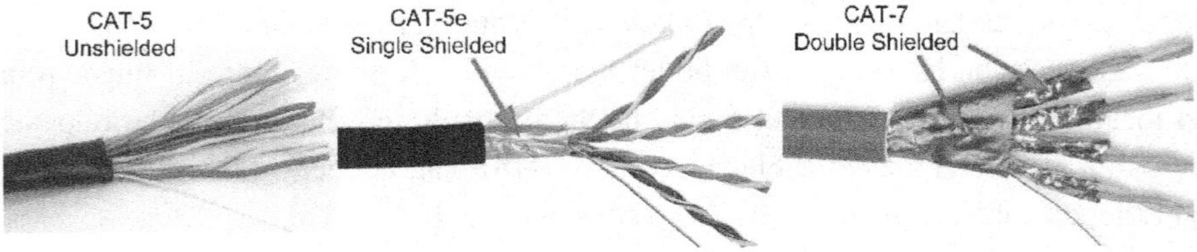

Figure 33 - Ethernet Cable Shielding Differences

If your home network is wired with CAT-5 cable, the good news is that you can use it to make these hard-wired network connections. The bad news is that it will not be as quiet as CAT-7 or CAT-8 network and should be upgraded at some point. Much like the weakest link in the chain, mixing CAT-5e devices with CAT-7 cables can improve the noise and speed of your home network but the best results are obtained when all connected devices are certified to CAT-7.

CAT-7 and CAT-8 networks also offer more speed which is a great time-saving feature for moving large files around, but it is their extra shielding that makes CAT-7 and CAT-8 networks more attractive for audio and video streaming use.

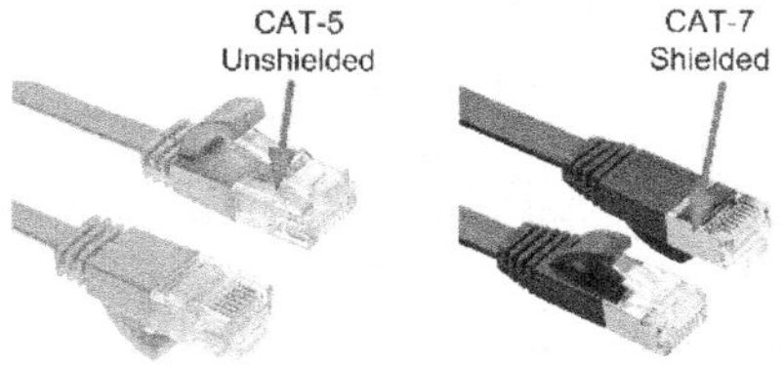

Figure 34 - Ethernet (RJ45) Connector Shielding Differences

The last issue in an NAS music server system is that Ethernet cables do not plug directly into each other (i.e., NAS directly to PC). Here you must have a router or switch to properly manage the network traffic. If your cable modem has extra ports on the back, it also contains a built-in router.

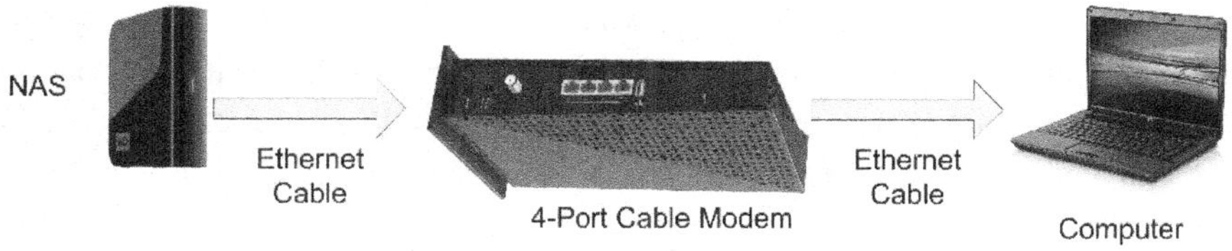

Figure 35 - Hard-Wired Router Connections

Some modems only have one Ethernet port on the back (no internal router). In this case, additional hardware is needed to make these connections. The picture below shows how to connect a single-port cable modem to a multi-port router.

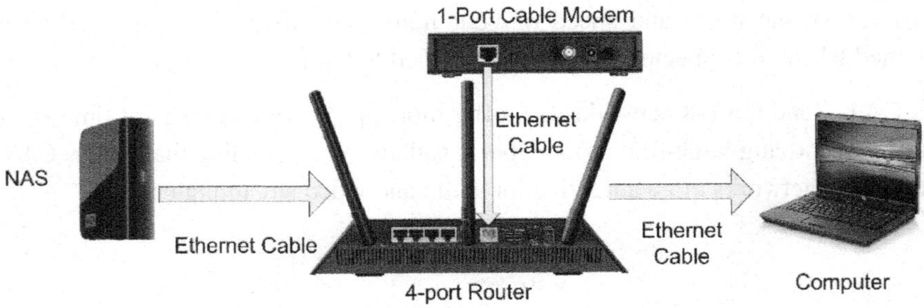

Figure 36 - Hard-Wired Modem+Router Connections

It is important to plug the Cable Modem into the WAN port (e.g., the yellow port) of the router. It does not matter what LAN port the NAS or Computer use.

The above network hardware configuration works well in most cases. Since the length of any single Ethernet cable in your LAN can be up to 300 feet, you can connect one long cable from the modem to the router and position the router, NAS, and Computer near each other to minimize the other Ethernet Cable lengths between them.

It is best to keep all Ethernet cables as short as possible instead of coiling them up in a pile. Also remember that abrupt bends or pinches to your Ethernet cables or roughly pulling them through blind spots and walls can damage its internal shielding. Be gentle with your cables and they will serve you well. Use common sense and never run your cables under heavy items where they could be squashed flat like under a piece of furniture. Crawls spaces, basements, attics and behind baseboards are all good options for cable runs.

Since the optimum location of your Wi-Fi router may not be in your listening room, you may prefer to locate your router somewhere central within your home so that it can provide a strong signal anywhere in your house. Instead of running two long Ethernet cables to the router, you may add a Smart Switch near the Computer and NAS to minimize wiring runs as shown below.

Features of a Smart Switch that are useful for home use are:

1. *Bandwidth limiting QoS per port* - you can limit devices with good granularity (i.e., you can stop you kids from hogging all the bandwidth)
2. *Pork trunking* - you can combine ports - useful if you have a NAS with twin network cards to improve reliability and increase network transfer speeds

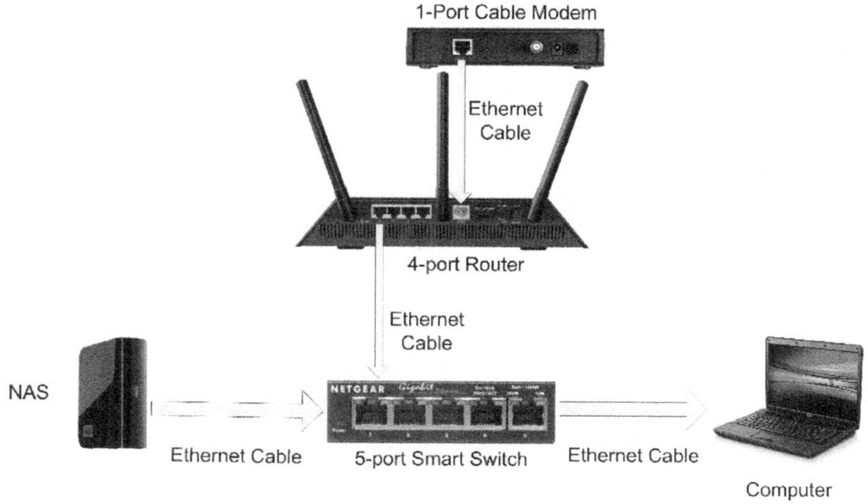

Figure 37 - Hard-Wired Smart Switch Connections

It does not matter which of the available ports you plug these three connections into on the switch; the switch will know what to do with all of them. The number of ports you buy on the switch also does not matter, just as long as there are more than two. I personally prefer using a *Smart* or *Managed* Ethernet Switch so that traffic on any hard-wired connection to this switch will be optimally controlled.

In summary, the biggest disadvantage of using a NAS over a DAS is its additional complexity and need for supporting equipment (Smart Switch, Router, etc.).

The biggest advantage of using a NAS in your streaming media system over a DAS is that your entire media collection can be accessed by every device you have authorized to connect to your LAN. Any authorized device can also remotely access your NAS anywhere in the world you find an internet connection. This may be important if you spend a lot of time away from home or if you wish to access your media from your second home. Remote access will not deliver the same quality as local access because of the bandwidth you purchased from your Internet Service Provider (ISP) as we shall see next.

Bandwidth vs. Throughput

Before we go too far down the rabbit hole, you need to understand the difference between two related data transfer rate terms: bandwidth and throughput. *Bandwidth* is the transfer speeds you pay for and receive from your Internet Service Provider

(ISP). Bandwidth is a measure of upload/download (*up/down*) speeds from your Wide Area Network (WAN). In other words, bandwidth determines how responsive data transfers occur to/from external sources on the Internet to your computer.

Upload bandwidth determines how fast data is sent from your computer to the Internet. *Download bandwidth* determines how fast data you request from the Internet gets sent to your computer. An ISP usually bundles bandwidth packages as 1/10 or 10/100 meaning 1.0 Mbps up and 10.0 Mbps down or 10 Mbps up and 100 Mbps down respectively.

Purchasing high bandwidth from your ISP may not be required, depending on what you want to do and what degradation you can expect during peak hours. For example, the recommended *minimum* Internet download bandwidth for playing a Netflix® video are 3.0 Mbps for SD, 5.0 Mbps for HD, and 25 Mbps for UHD quality. If you do not watch UHD, you may only need to buy a download speed of 5.0 Mbps although twice that would pretty much guarantee no dropouts, pauses, or spooling. On the other hand, if you want to stream UHD (*4K*) you must purchase a much higher bandwidth package.

Throughput is a measurement of data speeds between any two devices connected to your LAN. The same guidelines used for determining data transfer rates on a WAN also apply to the data transfer rates on your LAN. The throughput of your LAN must be equal to or greater than the bandwidth you purchase from your ISP. As described earlier, the throughput of your LAN is a function of your computer's network cards, routers, modems, and of course the Ethernet cables.

Remember that the weakest link in the chain scenario also comes into play when getting the bandwidth you purchase from your ISP through your LAN to your computer. It makes no sense to buy more bandwidth if your LAN cannot support it. See also Wi-Fi Bands and Speeds later in this guide.

All of the recommendations so far use the built-in components of the computer to process the digital signal and turn it into an analog signal. The part of a computer that does this is called a digital-to-analog converter or DAC for short. Stepping up to a DAC will dramatically improve any of the DAS or NAS systems you currently use as we shall see in the next chapter.

Digital-to-Analog Converters

The configurations described so far use the built-in digital-to-analog converter (DAC) of your computer. The output of its built-in sound card (mini-phono plug) connects to one of the inputs of the receiver (RCA jack) as shown below.

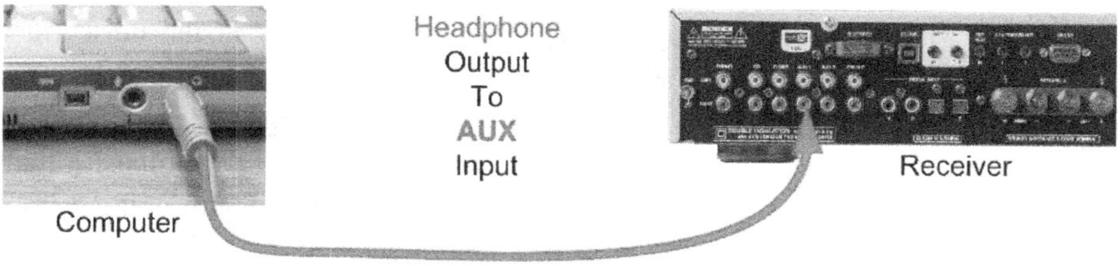

Figure 38 - Sound Card Output to Receiver Input

Wireless connections to a wireless A-V receiver are also possible thereby eliminating the need for this cable. However, I always recommend using hard-wired connections to reduce noise and improve fidelity.

In this configuration, any song you play with your computer's music player application will be heard through your receiver. While this is a very good starting point for listening to music, it is not of a configuration that supplies the best analog signal to your audio system. You are literally at the mercy of the computer's internal components as to the quality of sound that reaches your ears. Since most computer manufacturers are less concerned about how their computer sounds as they are about its computing power, this is a logical place to perform an upgrade.

You can eliminate the built-in constraints of your computer's internal components by adding an outboard DAC. Outboard DACs bring a whole new level of quality to the analog signal your receiver gets. The four problems typically addressed by outboard DACs include clock jitter, non-liner conversion errors, power supply, and output amplifier designs.

- *Clock jitter* in DACs translates audibly to minor changes in tone for sustained music. For example, when a long chord lingers in a performance, DACs with clock jitter issues will sound as if the melody contains a slight vibrato.

- *Non-linear conversion* issues in DACs translates audibly to minor changes in sound levels. For example, when a long chord lingers in a performance, DACs with non-linear conversion issues will sound a little louder or softer from moment to moment.
- *Power Supply* issues translate audibly to several different areas. The first area is in dynamic range where the power supply voltage actually sags during peak levels. This translates audibly to a compressed or even muted, shrill, or edgy sound. The second area is in the noise floor where the zero signal level is not a quiet as one would expect. This translates audibly to a low-level grunge changing the way a musical instrument or voice sounds. Hall ambience in live performances may sound constrained or undefined as echoes fade off into nothing. The third area is in loss of inner detailing. Related to dynamic range, this is the ability of a DAC to accurately reconstruct low-level harmonics of an instrument. This audibly translates to an alto saxophone sounding more like a tenor sax.
- *Output amplifier design* issues are also addressed. Typically consisting of one integrated circuit (IC), this output IC is literally the weakest link in the audio chain. The sonic qualities you get out of your computer are little or no better than the sonic quality you get from your cell phone. Much like adding an outboard amplifier improves the sonic quality of your car stereo, adding a better output amplifier to your computer does exactly the same thing.

Outboard DACs not only provide greatly improved fidelity over your computer's built-in DACs but they also decode more and different types of audio formats. For example, some built-in computer DACs may not recognize any type of DSD files while others may only recognize DSD64 files. Some outboard DACs even offer software and hardware upgrade paths for future growth.

Direct-Connected DAC

USB DACs are the most common type of outboard DAC on the market today. This is because they are so simple to add and requires essentially no extra knowledge or other equipment to make it work properly.

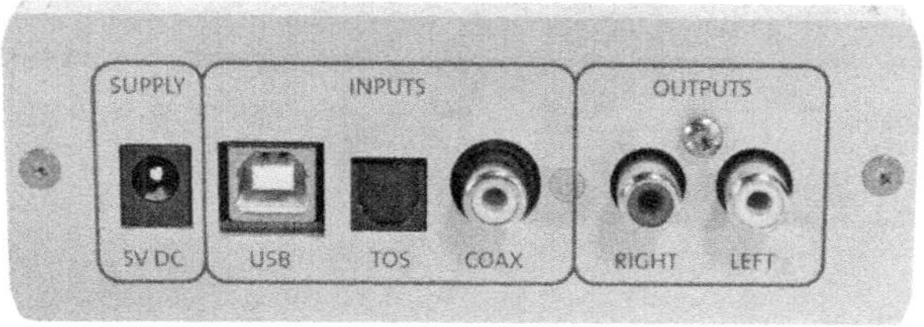

Figure 39 - A USB DAC Rear Panel Connections

An outboard DAC is directly connected to your computer via a USB cable (much like you would connect a USB printer to it). Instead of using the headphone jack and cabling shown in Figure 38, the output of a DAC is connected to the receiver with a different cable as shown below.

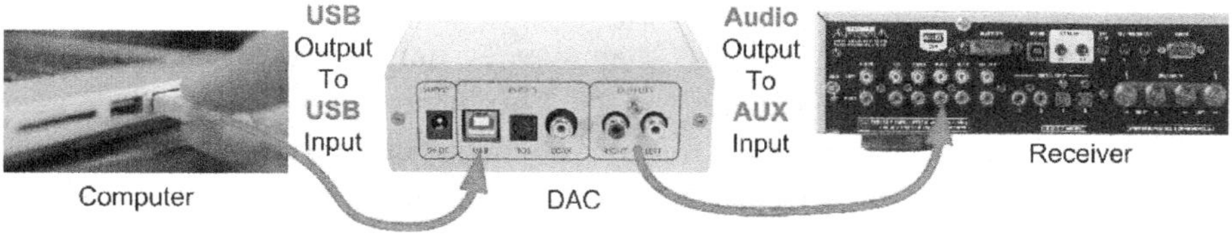

Figure 40 - A Directly-Connected USB DAC

The above picture shows an RCA cable connecting the DAC to a receiver but balanced outputs are also available from higher-priced DACS.

Although we have talked only about USB DACs as a direct-connected DAC, other types exist and just use different cabling. Instead of a USB cable, these direct-connected DACs use a TOSLINK™, FireWire, eSATA, or Thunderbolt cable between the computer and the DAC.

Ethernet DAC

An Ethernet DAC offers far more in the way of flexibility than any direct-connected DAC including remote connectivity. If you have cable-length issues between your computer, DAC, and receiver, an Ethernet DAC may be exactly what you need. The upside of an Ethernet DAC is that you can place this DAC right next to your receiver keeping the audio cable lengths very short. Adding an Ethernet jack nearby the DAC also allows you to locate your computer anywhere, or even wirelessly.

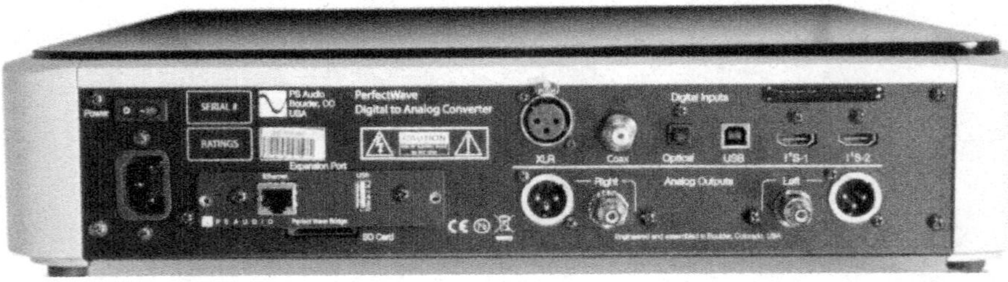

Figure 41 - An Ethernet DAC Rear Panel Connections

The downside of an Ethernet DAC is that they are more complicated to administer, require more hardware, and some Macintosh computers may not work well with them. Windows and Android computers, on the other hand, will love this type of device.

Unlike a direct-connected DAC that requires only a simple cable to your computer, an Ethernet DAC requires one more component in the link between your computer and your receiver called a *router*. An Ethernet cable is used instead of USB cable to make the connection between the computer and the DAC. Audio cables (RCA or balanced) are used between the DAC and the receiver as shown below.

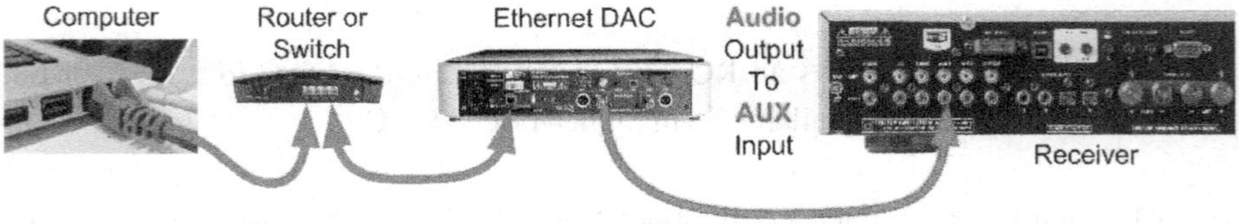

Figure 42 - Connecting An Ethernet DAC

In summary, the first sonic upgrade path up in migrating to a high-quality media server is to add either a USB or Ethernet DAC to your computer. The next upgrade path may be a larger external digital storage device. Any DAS or NAS device can be connected to your computer and used with either type of outboard DAC. But while we are again on the subject of digital storage devices, let us see what possible configurations exist.

Storage Devices

A computer disc (internal or external) stores the files (something you already know, right?). BUT when using only one disc, this single disc is vulnerable (one day, it will fail) and all your data on this disc will be lost. Most folks use backups for this reason but even these backup devices typically use only one disc and can they too can fail. The reason you use a backup is that the likelihood of *both* discs failing at the same time are very low. However, this also means that you must be diligent in doing backups and despite their best intentions, most people are not.

Automated backups assure you that a computer program watches out for your data and routinely does a backup for you at pre-determined intervals. This is good but it also means that your backup drive must be connected to your computer in order to do this at the time this automated backup is scheduled. If you use a laptop, this may not always be possible even with wireless connections. There are, however, other more fancy solutions as we shall see.

We can thank all of the failed hard drives in the past for the development of these fancy solutions; they are termed **RAID** storage systems. RAID stands for the *Redundant Array of Inexpensive Drives* and there are several configurations from which to choose, each assigned a numeric suffix. All RAID systems require a different hardware or software than normally provided in your computer.

One last note of caution regarding data integrity: Murphy's Law appropriately applies here where *anything that can go wrong, one day will.* In large computer centers the best designed servers are always backed up to another server. This extreme type of redundancy usually involves a remote location and regular automated backups made to that remote server. For this reason, most RAID storage systems are also bundled with automated backup software to back its own contents to another device (more hardware is required but more reliability is gained).

RAID storage systems also have a feature that can send you an email alert when a drive fails. This is exactly what you need to assure yourself that the integrity of your data is rock solid. When you get such a message, you must replace the failed drive immediately or again risk losing all of your data.

Regardless of RAID configuration type, you can think of any RAID storage system as one huge data drive. Let's look at two of the most common configurations for use in your media server: RAID-1 and RAID-5.

RAID-1

This is the entry point to a series of solutions that eliminates the need for performing automated backups. Here an internal controller uses two computer discs; whatever is stored on the first disc is also stored on the second at exactly the same time. This storage system is sometimes called a mirror since what you see on drive-0 can also be seen on drive-1.

The upside of a RAID-1 system is that if one drive fails (the most common failure mode), no data is lost. This is a good thing but the downside for this margin of safety is that it literally doubles the cost per megabyte stored (plus a small cost for the box to hold these two drives).

Figure 43 - A Two-Bay RAID-1

The other limiting factor in a RAID-1 configuration is its maximum capacity is limited to the largest single-drive size. In other words, you can use various sized drives in a RAID-1 system but its final size is equal to the smallest drive. For example, if you used a 4 Tb drive with a 3 Tb drive, the maximum capacity of the RAID-1 drive would be 3 Tb.

RAID-1 drives will still operate as if nothing was wrong with a failed drive in place. However, the data performance is less (data transfer is slower) since the RAID is operating in other than normal mode. If the second drive fails with an already bad drive in place, again all data will be lost. However, the chances of two drives failing at the same time are pretty small but it could happen.

I personally feel that a RAID-1 drive is the **best point** from which to begin building your media server. As of this writing, you can buy a 2x8 Tb RAID-1 system brand new for under $500. This gives you 8 Tb of storage with complete peace of mind. To me, this is a very reasonable price to pay for this level of insurance. Another RAID configuration is possible that provides additional insurance against multiple simultaneous drive failures as described next.

RAID-5

As mentioned for RAID-1 drives, the total capacity is limited to the largest drive used in the array. If you need *more* storage (beyond the largest single-drive size) you must use a different RAID configuration (or an additional RAID-1 drive). The next step up in an external drive for your media server is the RAID-5 configuration where a trick called "striping with parity" is used to store the data while also building-in very high reliability.

The upside of a RAID-5 system is that it can be quite large, depending upon how many drives there are in the array. The minimum number is three drives and the maximum is controller dependant (usually four to eight drives for home use). The total capacity of the array is one less than the number of drives in the array. For example, if you use four 3 Tb drives in a RAID-5 configuration, the capacity of the array is 9 Tb (not 12 Tb as you may expect).

As it was with RAID-1 drives, you can use various sized drives in a RAID-5 array but its final size is a multiple of the smallest sized drive. For example if you have five 4 Tb drives the capacity of the array is 16 Tb (not the 20 Tb that you may expect). The price paid for this level of insurance is the cost of one extra drive.

RAID-5 drives will still operate as if nothing was wrong with a failed drive in place. However, the data performance is less (data transfer is slower) since the array is operating in other than normal mode. If a second drive fails with an already bad drive in place, all data will be lost. The chances of two drives failing at the same time are pretty small but it could happen. If you wish to protect your data even under these circumstances, there is yet another solution, but the cost is one more drive.

This extra drive is called a "hot spare" meaning that it just sits idle in place waiting for one of the other drives in the array to fail. And when a failure is detected, the controller automatically removes the failed drive from the array and recruits the spare. The controller then rebuilds the array and disables the failed drive. If a second drive fails, no data is lost but the array performs at a reduced level.

Hot spare drives occupy a drive bay just like all of the other drives in the array. Most RAID systems use colored lights to indicate good (green) or bad (red) drives. One more feature – called hot swapping – allows you to replace a failed drive without having to shut down the RAID.

The capacity of the hot-swap array depends on the number and size of the drives. As it is with RAID-1 drives, you can use various sized drives in a hot-swap RAID-5 array but its final size is a multiple of the smallest sized drive (less two drives). For example if you have five 4 Tb drives the capacity of the hot-swap array is 12 Tb. Reduced storage capacity is the price of increased reliability.

Highly sophisticated RAID systems also use redundant power supplies and even multiple Ethernet connections to assure no data interruption in the vent of other types of failures. Wall Street uses such sophisticated systems because data loss cannot be tolerated. In such systems several hot spares can be reserved thereby preventing a catastrophe at the price of a second, third, or fourth hot-spare drive.

One can go crazy with redundancy where you can even backup one RAID-5 server with another at a different physical location. This means that in the event of a total loss from a fire or flood at one location (like your *summer* home), an ongoing automated backup from an identical server at another location (like your *winter* home) assures you that in such an unlikely catastrophe your data is not lost. In this way both servers have identical audio/video libraries at both locations but both servers must be identical in capacity.

As a side note, since RAID arrays can be huge, you can also combine its function as a media storage device with automated data backups of your computers. If you decide to do so, just add up the capacity of all the drives in your computers to the capacity you need for your audio/video files – plus normal growth. Here, if you have three computers each with 1 Tb drives and a known media capacity of 8 Tb, you would need about a 12 Tb RAID (8+1+1+1+growth). Adding multiple functions to your RAID server can also help you justify its higher cost.

Figure 44 - A Five-Bay RAID-5 System

RAID-5 systems are my personal favorites for building a media server. The downside is in the cost of multiple drives and the upside is low chance of data loss. In my mind, this is a very cheap insurance policy at the going price of disc drives today.

RAID storage systems are available with USB, Thunderbolt, Ethernet, Firewire, or Optical connection options. Whether you decide to use a hot-spare and/or the hot-swap capability is less important than moving to a RAID-5 system from a RAID-1 or single drive storage system. Which RAID system you buy is more of a risk vs. investment decision that only you can make. Regardless of how you choose, there is one more feature of a RAID system of which you should be aware: growth.

Growing the Capacity of a RAID-5

Most RAID systems can "grow" in size where if you buy 2 Tb drives for the array and decide to increase its capacity, you can do so without losing your data by installing one bigger drive at a time and instructing the software accordingly.

Let's assume that you buy a 4-bay RAID-5 system and populate it with four 2 Tb drives for an array capacity of 6 Tb. One day you see that your drive is reaching over 90% of its capacity (yikes!). Instead of buying a 5-bay RAID-5 system, you can add bigger drives to your 4-bay system. If you buy four new 8 Tb drives, you can increase its capacity to 24 Tb and all you must do is add one new drive at a time to the existing array.

A simple rebuild of the array after each drive addition moves the data from the old 2 Tb drive to the new 8 Tb drive. After the last drive is added and the array rebuilt, the capacity is still 6 Tb. In a second step, another command grows (resizes) the array to the new capacity making it a 24 Tb drive. No data is lost during this operation.

Hard Drive Types

Most people understand that a hard drive is inside of a computer and it stores long-term digital information. But the hard drives inside home computers are designed and built differently from those in RAID systems. Manufacturers build drives to a failure-rate specification called the Mean Time between Failure (MTBF). Higher reliability is incrementally more expensive with higher MTBF numbers.

But like the worst thing you can do in your car is to start its engine (oil drains from parts when turned off and highest frictions occur when starting up), a similar thing happens for all electronics when turning them on. When you turn on your computer, micro power surges momentarily race through before things settle down and it is these tiny split-second surges that typically cause failures. Fortunately these surges *usually* do not damage anything but they do impact the MTBF figure.

Building a reliable hard drive addresses this and other mechanical and electronic issues and in doing so increases the selling price of the drive. Looking at the various types of internal hard drives manufactured by Western Digital, they produce six basic categories focusing on reliability, performance, and capacity: Blue, Black, Green, Red, Red Pro, Purple, and Gold series. Looking at similar hard drives manufactured by Seagate, they produce the Barracuda, FireCuda, IronWolf, IronWolf Pro, and Skyhawk series. So which one should you choose?

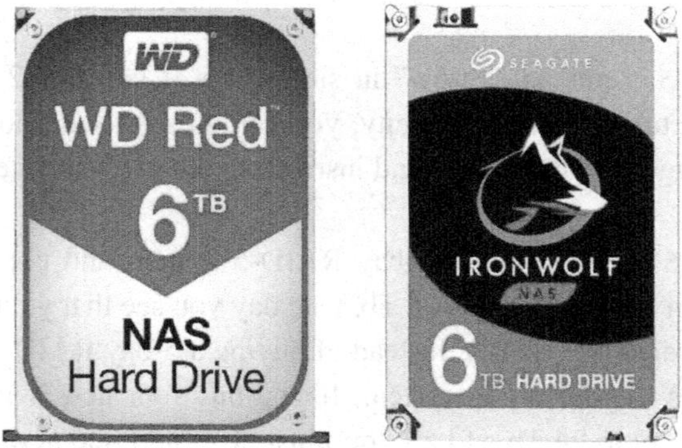

Figure 45 - Good NAS Hard Drives Are Essential

That's a really good question and fortunately drive manufacturers have on-line and chat-based solutions to help you find the perfect drive for your individual needs. If you wish to use your DAS/NAS to stream video as well as audio, you may want to focus more on performance and capacity. If you wish to use your DAS/NAS to only stream audio, you may want to focus more on reliability and capacity. If you are concerned about everything, you may wish to focus on self-healing features.

The choice boils down to cost vs. return where whatever you choose should give you the most peace of mind. Ask yourself this question: "Is it really worth spending more for a better drive?" Then ask yourself this follow-up question, "Am I being penny wise and pound foolish?" These are questions only you can answer. I personally use the Western Digital Red drives (not the Red Pro) in my 4-bay NAS. These drives offer a good tradeoff between cost, performance, and reliability.

Summary

To summarize your options for selecting the storage device for your media server, you can choose between a direct-connected DAS and a remotely-connected NAS. The hard drives inside the DAS or NAS should be of a high-quality type so as to assure you of a long, reliable life. Reliability translates directly to risk of loss of your entire audio/video collection.

Putting all of your data on one drive is *very* risky and using a backup scheme assures you that at the least another copy of your data can be retrieved in the event of a catastrophe. Backup copies can take many forms but all involve inconvenience of some sort whether it be performing a manual backup or swapping out one drive for another or adding more cost for automated backups.

RAID servers offer a compromise where built-in redundancy automates this backup necessity. RAID servers must also be backed up but are highly reliable and very low risk. RAID servers can be either DAS or NAS.

Figure 46 - DAS and NAS Storage Solutions

The error recovery controls built into NAS-class and Enterprise-class hard drives are worth their premium prices if for nothing else to give you the peace of mind that your library is as reliable as possible. These classes of drives offer data protection features that minimize data corruption under normal wear & tear and in the event of an unexpected power loss.

Before we go on, let's discuss the various power issues in your system, how important this is to the integrity of your audio/video library, and how it can even help improve your system's audio and video fidelity.

Power

At this point in building your media server, you have some sort of internal or external drive connected to your computer. This drive, regardless of its type or configuration, is susceptible to another type of catastrophic failure: power loss. We will cover this issue and a few other power-related issues in this chapter.

The second-worst thing you can do to a computer is to unplug it while it is running (that is, to perform an unplanned shutdown). Computers are constantly performing background tasks and merely flipping the power switch can produce catastrophic results. The safest way to turn off a computer is to perform a deliberate and safe shutdown (e.g., START > POWER > SHUT DOWN).

Figure 47 - Loss of Computer Power Causes Catastrophic Results

But safely shutting down your computer may not always be possible. Random power outages routinely occur from your local electrical company as they perform regular scheduled (or unscheduled) maintenance. Normally, these interruptions in service last a few moments and rarely cause problems. But compared to appliances, computers are another story where they cannot tolerate even minor disruptions in power (laptops with good, charged batteries are an exception).

UPS

The next item you need in your building your media server is a really good *uninterruptable power supply* (UPS). Even if you use a laptop as a media server, a UPS is a wise investment to – if nothing else –preserve the integrity of your stored media on your external DAS or NAS.

It is not necessary to have enough power from your UPS to run your entire home theater, although some folks do and you may wish to do the same. At a minimum, two essential items that must be powered by the UPS: your **computer** and your **external storage** device. A *smart UPS* has the ability to signal your computer via a USB connection to perform a safe shutdown in the event it cannot maintain power beyond the length of the power loss.

The type of UPS you need is called a *pure sine wave* UPS. This type of UPS delivers the same type of electrical power to your delicate electronics that your wall outlet does, and sometimes even better. Other types of UPS designs – like modified sine wave or square wave inverters – are lower in cost but do not provide the same type of power as your wall outlet and typically you will hear noise in audio tracks and/or see edgy artifacts in video streams.

The amount of power a UPS provides your system (capacity) is the other factor you must consider during its purchase. The way you do this is simple: add up all of the number of Watts listed on the data labels on the backs of each unit and then double it. This will tell you the safe number of VA (volt-amps) you require from your UPS. For example if your PC draws 150 Watts and your DAS draws 50 Watts, the smallest UPS you should buy should deliver 400VA (twice the wattage draw). Although this is not technically correct, it is a simple way of figuring out how much power your UPS *should* deliver without getting into more complicated calculations.

So now you know the capacity in VA of your UPS. The next thing you want to determine is “how long should the UPS hold up your system?” This is entirely up to you. I would recommend at least 15 minutes since the vast majority of power outages are shorter than this. But if you live in an area that regularly experiences prolonged power outages, you may want to buy a longer-life UPS.

The total time any UPS can hold up its rated capacity is called its *run time.* Longer run times literally translate to bigger sizes. A small physical sized UPS with one battery cannot provide backup electricity as long as a larger UPS with several internal or even external batteries. The number in the specifications of a UPS that hints at its run time is the total number of Amp Hours (Ah) of all batteries. Here, the larger the number of Ah, the longer the run time. For this reason, some UPS manufacturers offer external batteries to provide extended run times.

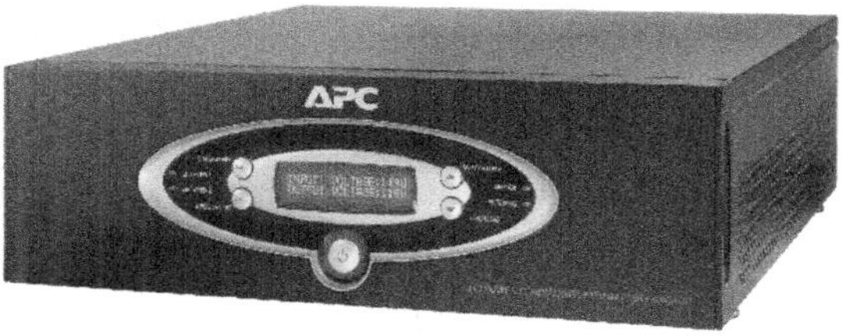

Figure 48 - A High-Quality UPS with a Long Run Time

Some UPS units also have the ability to remove unwanted noise that normally comes from your wall outlet. These units have special filters called *line conditioners* (*power conditioners*). However, there are also specialized outboard line conditioners that can work even better than those built into a UPS.

Line Conditioners

The purpose of a line conditioner is to remove the unwanted noise (called *dirty power*) that exists in the electrical power coming from your wall outlet (*AC Mains*).

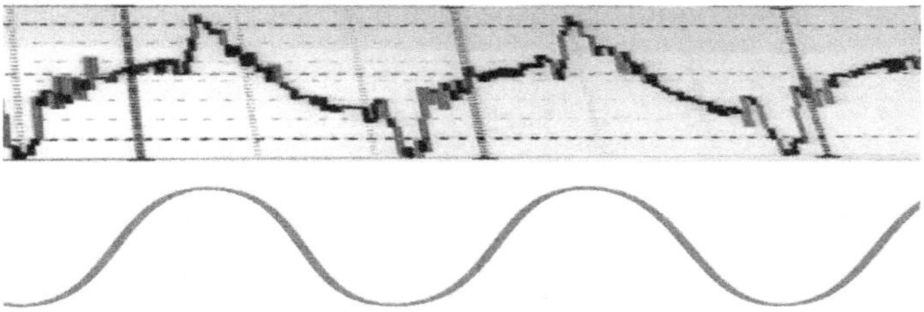

Figure 49 - Dirty Power (Top) vs. Clean Power (Bottom)

Dirty power adversely affects *every* aspect of your media server. In your audio system dirty power can diminish inner detailing, front-to-back dimensionality, height-and-width of the sound stage, and low-level background noise. In your video system, dirty power can diminish video resolution along with lifeless colors and poor contrast.

Figure 50 - A High-Quality Line Conditioner

Unfortunately, cleaning up this noise at your wall outlet or out of your UPS may not be a simple task. The amount of electrical noise you experience at your home depends on where you are along the distribution line with regards to other homes and businesses along the power distribution infrastructure. If your home is at the end of the line, you will see significantly more electrical noise than if your home is at the beginning of the line.

There is no way to know where along the distribution line your home is situated and there is no way to know how much noise other homes and industrial businesses are contributing to your dirty power problem. If you live near an industrial complex with heavy machinery, it is almost a certainty that the power coming out of your wall outlet is quite dirty. Regardless of your location, it is always a good thing to protect your valuable equipment with a good line conditioner.

Some line conditioning companies emphasize their ability to withstand surges and lightning strikes as the key factor in cleaning up dirty power. This is only partially true. While filtering out random spikes and sporadic power surges is a very good thing to do, it is not the only thing that must be done to clean up AC Mains noise. Surge protection is a different field of study than noise elimination and those units claiming to do both will in fact do so, but they may not do both well.

Surge Protectors

Like buying a separate preamplifier and power amplifier instead of buying a receiver, a dedicated design for a specific need usually creates a better product than one that tries to do everything in one box. It only makes sense that if a receiver and an amplifier cost the same, the sound from the amplifier will be very different from the sound of the receiver. In surge protection, this analogy also holds true.

In the picture at the right, you can see the dirty power as shown in Figure 49, but here you can also see an anomalous spike. It is these transient spikes that surge protectors are designed to eliminate. The circuitry used to address each type of AC Mains interference is different as is the science behind their designs.

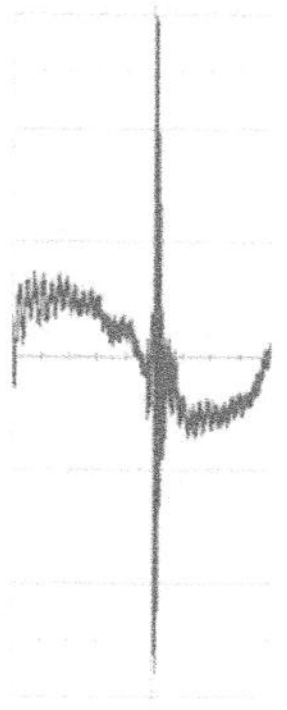

Figure 51 - An AC Mains Surge

Your electronics will operate just fine without addressing dirty power but to protect your investment, you should *always* have a surge protector on your delicate equipment. Adding surge protection with line conditioning allows you to experience your system's full potential.

In summary, choose a pure-sine wave UPS whose capacity (in VA) is twice that of the sum of the connected equipment (in Watts). The run time at this capacity should be at least 15 minutes but can be longer in areas where long power outages are routinely experienced. External battery packs can add to its run time. Features in a UPS such as surge protection and line conditioning may be useful but additional devices to provide these features may offer better protection and cleaner power.

While we are on the topic of power, let's briefly look into the design of the power supplies of your equipment since this topic of noise is common to both.

Power Supply Upgrades

When you plug a device into your wall outlet and flip on the switch, magic occurs inside that transforms the AC Mains power into music and video. How it does this is unimportant but knowing that you may be able to improve what you hear and see is worthy of mention.

There are two basic designs for converting the AC Mains power into that used by your devices: an *analog* power supply and a *digital* power supply. Both "work" but audiophiles/videophiles usually prefer the sound improvement heard and images seen from high-quality analog power supplies. Much like the dirty power filtered out by a high-quality line conditioner, a digital power supply generates its own dirty power. This dirty power is created downstream of your line conditioner so your line conditioner cannot remove it.

Since your media server becomes part of the audio/video chain, it can contribute noise into the audio/video signals from its internal digital power supply. Fortunately there are third-party manufacturers that address this issue who offer analog power supply upgrades for some digital devices.

For example, one of the most popular computers chosen for use in our local audio club is Apple's Mac Mini. This is a fine computer for DAS configurations with one of the smallest small footprints available. And for those of you who wish to look at less clutter, it can easily be hidden behind other equipment. But the stock digital power supply in the Mac Mini is quite noisy and the power to the built-in USB and Thunderbolt ports is "mini" as well (underpowered).

Because of its popularity, several folks in the industry have stepped up and designed far better analog power supplies eliminating these minor annoyances. As you can see from the image below, some analog power supply upgrades are no longer mini in size.

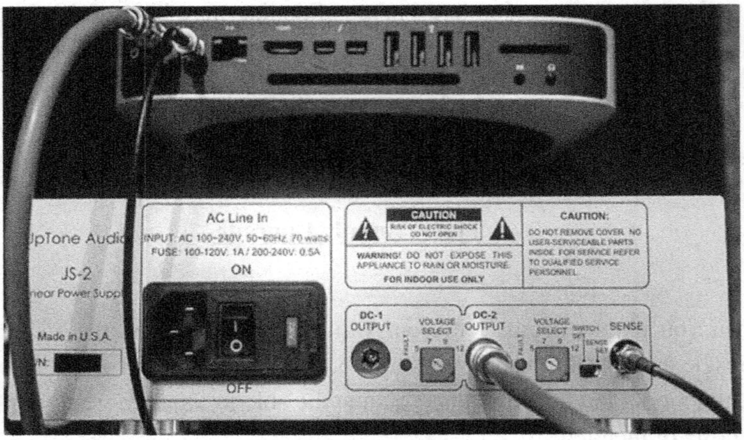

Figure 52 - A Mac Mini Power Supply Upgrade

When using a computer for surfing the web, word processing, or number crunching, standard digital power supplies work fine. But when stepping up to high quality audio/video, there is more to consider about the intrinsic noise digital power supplies introduce to your streaming media system.

The point of mentioning the Mac Mini is not to pick on Macintosh products but to show you that high-quality power is extremely important once you begin to step up in audio and video quality. Similar analog power supplies are available for a wide range of Windows and Linux products that provide the same improvements in fidelity.

Summary

In summary, choose a pure-sine wave UPS whose capacity (in VA) is twice that of the sum of the connected equipment (in Watts). The run time at this capacity should be at least 15 minutes but can be longer in areas where long power outages are routinely experienced. External battery packs can add to its run time.

Power from your utility company can inject noise into your home entertainment system. To eliminate this noise, you can buy devices to not only clean up this noise (line conditioner) but also protect against lightning and power surges (surge protector).

Removing electrical noise downstream of your line conditioners and lightning arrestors is possible by upgrading the digital power supplies to high-quality analog power supplies.

In the next chapter, we will discuss the upsides and downsides of wireless and wired networks and what you need to understand about these two approaches when it comes to audio and video fidelity.

Networking

Computer networking is a complicated issue and there are many things you need to understand before moving on to other detailed specifications about your media server. I must digress here at length so you can grasp the impact of these issues not only around Wi-Fi and hard-wired network connections but also other network-related issues that can impact audio and video fidelity. Let's begin with where most of you will begin and that is by using wireless connections.

Wi-Fi Connections

Wi-Fi is an incredibly convenient way to add your computer to your home network. But not all Wi-Fi computers are created equal, especially when it comes to the fast Wi-Fi speeds required to stream HD and UHD video. There are two component parts to any Wi-Fi connection: the computer-end and the router-end.

If you have a fast computer and a slow Wi-Fi router, you are at the mercy of the weakest link in the chain analogy once again. For example, if you have a gigabit router and a 100 Mb Wi-Fi card in your computer, your speed is limited to 100 Mb.

Think of this like download speeds when previewing movie trailers on your smart phone. With a good Wi-Fi connection, you get great video streams without interruption and conversely with a poor connection you get pauses in the video (the dreaded *SPOOLING* prompt). What this means is that the Wi-Fi's data transfer rate cannot keep up with the demand requested by your smart phone.

The same thing can happen in your home wireless network, especially if there are multiple Wi-Fi users on it when you wish to listen to stream high-resolution audio or video. Building in strategies now to handle future growth and demand is always a good idea. If you plan to use Wi-Fi connections, let's see what can be done to optimize this choice.

Wi-Fi Bands and Speeds

The two Wi-Fi bands used in the consumer electronics are 2.4 GHz and 5.0 GHz. The 5.0 GHz band typically offers faster data transfer rates but devices must be fairly close to your Wi-Fi router. Devices using the 2.4 GHz band can be operated further from the Wi-Fi router but the further away from the router, the slower the data transfer rates can be. Airborne interference (e.g., nearby cell towers, radio/TV transmitters, industrial parks, etc.) also slows down Wi-Fi up/down transfer rates on both bands.

The Wi-Fi components inside both your computer and your Wi-Fi router limit the data transfer speed you will experience on these two bands. These components are designed to five industry standards industry are 802.11a/b/g/n/ac. When both the computer and Wi-Fi router use the same standard (e.g., 802.11n), the speeds are optimized to this band. If one device uses an inferior standard (e.g., 802.11n and 802.11ac), the speeds are limited to the slowest standard.

Theoretical vs. real-life download speeds on these two bands for a single user and a dual-band Wi-Fi router are:

Standard	Band	Theoretical	Typical
802.11b	2.4 GHz	11 Mbps	2-3 Mbps
802.11a	5.0 GHz	54 Mbps	20 Mbps
802.11g	2.4 GHz	54 Mbps	20 Mbps
802.11n	Both	600 Mbps	40-50 Mbps
802.11ac	5.0 GHz	1300+Mbps	70-200+Mbps

All Wi-Fi components in your home use these same two wireless bands including your home entertainment system, cell phones, computer systems and tablets, and most other mobile devices. Because of this, *congestion* on your home Wi-Fi network may be heavier at one time than at another, especially when there are multiple users. Even with a robust Wi-Fi network in place, airborne interference and Wi-Fi congestion can cause momentary delays in audio/video data transfer.

There are basically four solutions to this problem. The first addresses the type of Wi-Fi router itself, the second assigns priorities to those devices connected to your Wi-Fi, the third uses hard-wired connections, and the fourth uses a second, completely separate network on which only your media devices are connected. All four solutions have their up sides and down sides as described next.

Wi-Fi Router Speed

When using wireless connections with your media server, you definitely want to use the highest connection speeds possible to minimize delays or dropouts. There are two types of wireless routers on the market today: dual-band and tri-band.

Dual-band routers with simultaneous technology broadcast two signals at once: a 5.0 GHz signal and a 2.4 GHz signal. Wi-Fi devices that support the 5.0 GHz band automatically connect to the faster 5.0 GHz signal, while any older devices automatically connect to the 2.4 GHz signal. So a dual-band router essentially hosts two different Wi-Fi networks at once.

Tri-band Wi-Fi routers take this approach one step further and provide three wireless bands: one 2.4 GHz signal and *two* 5.0 GHz signals. Currently, wireless AC3200 *tri-band* Wi-Fi routers offer the highest download speeds. Tri-band routers are your best choice for handling multiple simultaneous connections to your wireless network.

Figure 53 - Tri-Band Wirelss Routers Offer the Best Multi-User Performance

Although these tri-band routers do not guarantee your sons or daughters will not interfere with your streaming high-resolution content while playing their online video games, it is your best option for a wireless network. But there may be a solution to this potential congestion.

Wi-Fi Priority Management

On some Wi-Fi routers you can administer priorities for connected devices. This assigned priority ranks data transfer rates to those listed devices in the order of appearance. By adding your Wi-Fi connected media devices to this list you can better control drop-outs and pauses.

For example, in Figure 57, the media server named OPPO is at the top of the list and therefore assigned the highest priority of use. Any other connected Wi-Fi device yields data transfer requests to the item at the top of the list. Also shown are two other devices on this list are also granted higher priority rights than any other wireless device not on this list, although still subservient to those above it on this list.

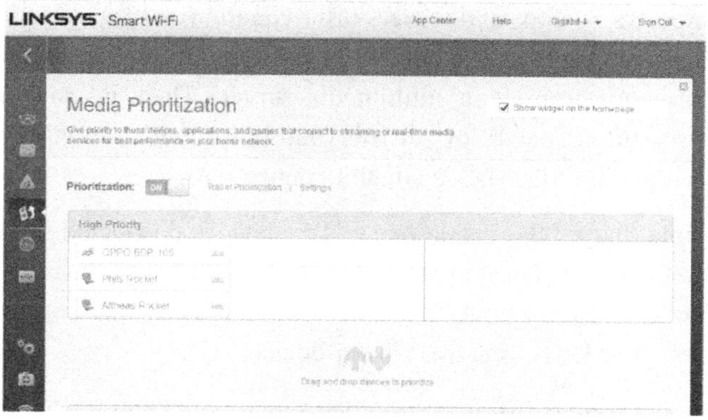

Figure 54 - Managing Priorities in Wi-Fi Networks

In assigning priorities, you must consider the impact it may have on the other users of your network. Throttling use by assigning priorities may make you happy but may frustrate other users.

Two other options (and in my opinion better options) involves a hard-wired Ethernet connection with your multimedia server or a second network for your streaming media devices. We will discuss these in more detail after we briefly discuss how network cards play a role in throttling back your LAN speed.

Hard-Wired Connections

On the rear panel of most Wi-Fi routers are several hard-wired Ethernet ports. These ports usually operate independently from that of the Wi-Fi network essentially giving you a third network (*switched network*). Using hard-wired connections is a bit more cumbersome but the benefits are huge.

As mentioned earlier when describing Bandwidth vs. Throughput, there is a difference between *internet* (WAN) speeds and *local area network* (LAN) speeds. LAN speeds are limited by the maximum speeds of each device used to manage the data transferred on your LAN (routers, switches, and hubs). If all devices are matched to the same high speed, all connected devices on your LAN will have the potential to transfer data at this high speed, even when using a WAN speed that is not as high. However, when mixing LAN devices of different speeds, data transfer rates are always throttled back to the slowest transfer rate through that LAN device (weakest link in the chain again).

Most computers, laptops and servers offer Gigabit (or 1,000 Mb) or higher hard-wired network throughput speeds and it is essential to use the highest speed possible for best performance from your multimedia server. Their internal network cards (NICs) typically offer speeds of 10/100/1000 Mbps which translate to potential network speeds of 1000 Mbs (i.e., a Gigabit connection).

It is possible that a NIC inside one of the devices connected to your LAN is a 10/100. This means that data transfer rates through this device has a maximum of 100 Mbps. While you may think that this is not an issue, this depends on how all other devices on your LAN treat this slower device.

Most Ethernet routers use what is called "smart switching technology" meaning that if a 10/100 NIC is present with a 10/100/1000 NIC, it will treat them as two different connect speeds switching the higher speed at its higher rate and the lower speed at its lower rate. Routers without smart switching technology will throttle back all devices connected to it to the slowest NIC speed. When buying any other device (switches or hubs) for your hard-wired LAN, verify the NIC speed of that device from the specifications available on the manufacturer's web site also provide a 1000 Mb (1 Gb) data transfer rate.

Multiple Networks

Possibly the best solution you could implement for your streaming media server is to add a dedicated network for use only by your media devices. The upside to this approach is that it electronically isolates all other network activity; non-media traffic and associated congestion is no longer an issue. The downside is that it requires twice the investment in your home network's infrastructure.

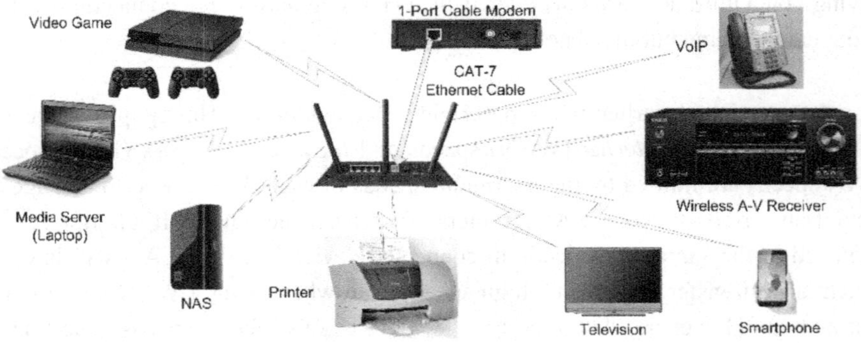

Figure 55 - Everything on One Wi-Fi Network

You should also consider *not* connecting your media network to the Internet. Not only does this eliminate non-essential network traffic but such isolation also assures you that there is no way pirates can hack into your system with viruses, Trojans, or even ransomware (more insurance).

Figure 56 - Use a Second Dedicated Network

To purists, no Internet connection is a very small price to pay considering its benefits. Plus any associated noise or distortion that could result from any kind of digital-signal interference is now completely eliminated by this isolation technique. And if your media-streaming network is also hard-wired with CAT-7 cable and components, any airborne interference is also minimized.

To give you the best of both worlds you can still stream from the internet through your Smart TV or other set-top device and still connect the audio outputs to your A-V receiver. Of course you must select the correct audio input to hear such streaming but again this is a small price to pay for total isolation.

You can still download and/or transfer files to grow your library via a thumb drive from another computer that *is* connected to the Internet. In this way, your anti-virus/anti-malware/anti-ransomware applications on your Internet-connected computer can screen these files before moving them onto your media network.

> Remember to always scan any files you intend to place in your library with your anti-virus/anti-malware/anti-ransomware applications before you attempt to copy, move, or use them.

In my humble opinion, a second network is one of the cheapest forms of insurance you can use to safeguard your media library and one that I would highly recommend you place somewhere on your upgrade list.

You probably know more now about wireless and wired networks than you really wanted. However, the importance of understanding how these network devices all synergistically fit together keeps you from buying the wrong configuration.

Selecting good devices for your WAN/LAN infrastructure impacts what you should look for in a computer and how you can expect it to perform. As long as the specifications for the network components are of high speed and quality, then you are reasonably assured that you will experience the best possible fidelity.

Lastly is the Ethernet cables themselves. As mentioned earlier, most existing LANs run CAT-5 cabling which is adequate for data transfer but does not provide noise immunity essential for high-quality audio/video streaming. For best results, use CAT-7 cabling at a minimum to all hard-wired streaming media devices.

Hopefully at this point you have a fair understanding of the possible network configurations and the upside and downside of each. Now would be a good time to discuss the specifications for the computer you wish to use as the heart of your media server.

Media Server Specifications

At this point of building your own media server, you have selected the type of computer (an existing computer or laptop), where you will store your library (internal or external DAS/NAS drive), and what type of network you will run (wireless, wired, or two networks). You may also have connected everything to a high-quality UPS and even have your sights on a DAC. This is a great starting point for your hardware needs and will get you into high-quality streaming media.

However, if you do not have an existing computer or laptop or if you wish to buy a new dedicated computer for this purpose, what should you get? This is another great question! There is a lot of snake oil, sales hype, or spin out there touting this feature over that but this guide does not subscribe to plunging into the high-end computer market as a first option. Instead, I encourage you to experiment with what you have even if it is currently your only computer. Once you are bitten by the media server bug, you will have grown in knowledge and experience and be able to understand what is and what is not important for you.

You may also wish to explore other user's experiences with your chosen solution and see if it will do everything you want it to do. Taking a few extra minutes and read or view reviews to help you from making a huge mistake.

For those of you who do not wish to assemble a system from scratch, at the end of this chapter I will discuss off-the-shelf (plug-and-play) media servers that can help you make a better-informed choice.

When buying a computer to be used as a media server, processing power is not as much of an issue as the other components in the system as we shall now see.

Computer Configurations

As mentioned earlier, it really does not matter what operating system your computer uses (Mac, Windows, Android, Linux, etc.) in putting together your media server. But the individual hardware components used to make up this computer does matter and in some cases they matter quite a bit. One may suspect that processing power of a computer (rated in Millions of Instructions per Second or MIPS) is a big concern, but for streaming media this is not that big of an issue. Here are some pointers to remember when picking out the internal hardware configuration for your media server computer.

Before we begin, there is one clarification I wish to make regarding the type of operating system used by the computer you choose and that is with the compatibility of the controlling software you also choose (JRiver, iTunes, etc.). Remember this:

Whatever software you choose to manage your library, it ***must be compatible with the operating system*** *of your computer.*

Fortunately, most major media server software applications (*Management Software*) run on a wide variety of operating systems but it's always best to check with the software manufacturer for compatibility with the computer's operating system you are considering before you buy. A partial list of Management Software appears in a later chapter along with the operating systems on which they can run.

As you may know, some manufacturers are building products that claim to also include the functionality of a media server (such as the SONY® PlayStation® 4 video gaming device). But what is it about the PS4™ that causes it to do both? Let's take a peek "under the hood" at the individual component differences between the PS4 and one of the SONY flagship computers to see the truth.

Figure 57 - Looking Inside Reveals the Truth

To give you an idea of the computing requirements for your media server, the table below compares the hardware specifications of the SONY® PlayStation® 4 to a high-end SONY VAIO™ laptop (same manufacturer, different design goals).

While it is interesting to note these obvious hardware differences, they are not the *only* differences between the two devices. For one example, the PS4 has a built-in optical drive while the VAIO Z does not. Power supply capacities, cooling, maximum numbers of hard drives are others but let us ignore these hardware differences for the moment and look only at the specifications.

Item	PS4	VAIO Z
Processor	AMD® Jaguar	Intel® i5-6267U
Clock Speed	2.13 GHz	3.30 GHz
Number of Cores	8	2
L2 Cache	16 Mb	4 Mb
Graphics	AMD Radeon Polaris	Intel Iris™ 550
RAM	8 GB GDDR5 2750 MHz	8 GB DDR3 1866 MHz
Storage	1 TB	256 GB SSD
Optical Drive	Blu-ray	None
Display	HDMI 2.0	HDMI 2.0
Audio	Digital output (optical)	Mini stereo
USB 3.0 Ports	3	2
Gigabit Ethernet	1	1
Wi-Fi	802.11a/b/g/n/ac	802.11a/b/g/n/ac
Bluetooth®	4.0 + LE	4.1

Although the PS4 is a superb interactive gaming console whose computer emphasizes processing power and graphics speed, it is still a media server by definition. But *which* of these specifications also makes it a media server? What *is* the truth before the "marketing hype," the "snake oil," or the "spin?"

Getting back to the goal of computer specifications, let's begin with some basic assumptions. Below is a list of hardware specifications you should consider when picking out a specific laptop or desktop computer to be used as the computer for your media server. Remember some physical design differences are not covered.

Assumptions

1. You have decided to **build your own media server**.
2. You want to use the less-complicated method of **wireless** connections.
3. You wish to stream both **audio and video** files.
4. You have decided to start with **internal storage** and then move up to a DAS or a NAS at some time in the future.
5. You want to **buy a new computer** specifically dedicated (not in shared) for this purpose.
6. You are **undecided** on which to buy: a laptop or desktop computer.
7. You are **uncertain** about what to look for in the way of features or options and you do not want to buy more than – or overlook – something that you need.

The great thing about starting with this approach is its flexibility and ease of expandability. You can start with an internal drive, upgrade you media server in steps while listening to and becoming familiar with its characteristic sound, and appreciate each new sonic improvement as you perform every upgrade. This approach also spreads out the cost of your investment over time allowing you to get into high-quality streaming media and out of compressed audio and video.

Processor Speed and L2 Cache

Faster processors and multiple processors (*cores*) with more threads and more on-chip L2 cache will respond to commands quicker than those with slower single processors, fewer threads, and less on-chip L2 cache. For media servers, this is important since video has a much higher data stream (bandwidth) than that of an audio stream. And if you are entertaining the idea of storing Ultra-High Definition (4k) video on your DAS/NAS in the future, you should probably buy the best performance processor you can afford.

Figure 58 - Processor Speed Is More Important For Video Servers

On-Board RAM

On-board RAM capacity and speed is another area where improvements can be found. More RAM with faster clock speeds allow the computer's operating system to perform more responsively. More RAM allows adequate storage space for temporary operating system, shared video, and the audio/video file to be played. Higher RAM clock speeds and dedicated video card memory (as opposed to shared video memory) also increases computing power since RAM access times are lower (faster) compared to hard drive access times (slower). Faster access times translate to a quicker response from one of your requests.

Figure 59 - More RAM Allows The OS To Be More Responsive

Most audiophiles prefer the sound of streaming from a music file that is buffered (temporarily stored) in RAM. Here, capacity is more important than speed so higher capacity RAM is more desirable than slower speed RAM (although both high capacity and high speed are best).

This would be a good time to summarize everything about the specifications that make up a good media server laptop or desktop computer. There will be more about optional accessories you can add on to either covered later in this guide but here is where we are at right now.

Laptop Specifications

First, a laptop is an excellent good choice for a media server since it already has many things bundled into it, such as a monitor, keyboard, mouse, and backup power supply. The other features of a laptop as a media server are that it is very small and consumes very little power. But there are numerous laptop configurations and at a minimum, these are the things you need when shopping for your laptop.

Item	Specification
Wi-Fi Card	Dual-band
NIC	10/100/1000
Wi-Fi Router	Tri-band
Processor	2GHz or more
Number of Processors	2 or more
RAM	8 Gb or more
Internal Hard Drive	1 Tb or more

Figure 60 - Laptop Minimum Specification

The supporting hardware components that you need for this approach are:

Component	Specification
Wi-Fi Router	Tri-band
Monitor	No
Keyboard	No
Mouse	No
UPS	Maybe

Figure 61 - Laptop Supporting Components

For laptops, there is no need to buy a high-resolution screen, touch screen, or separate video card since this computer will be dedicated as your media server. While a 1 TB drive at a minimum is recommended, remember that high-resolution audio/video files are very large and you may quickly run out of room. This is why a DAS system is recommended over an internal-drive system.

Below is a picture of the required laptop connections using a wireless A-V receiver. As you can see, this is a minimalists approach to get started in building your media server.

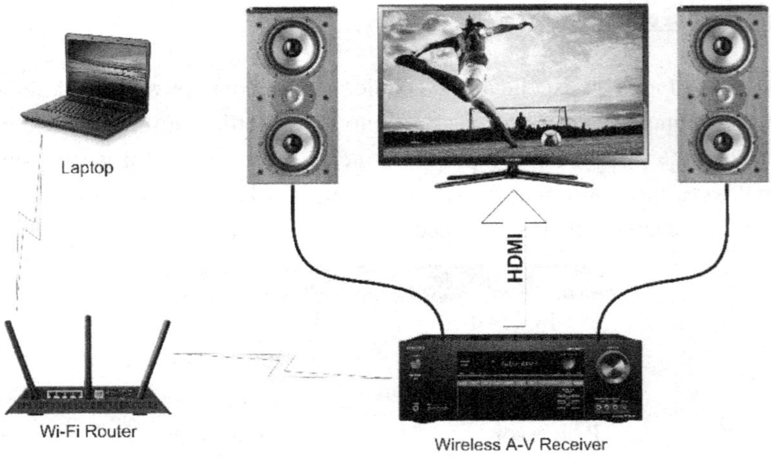

Figure 62 - Wireless Laptop Connections

The only real downside in using the internal hard drive of your laptop is in its limited size. However, its simplicity will get you started down this path and for many of you I imagine this is how you will begin. Remember adding capacity just means adding an external DAS through an available USB port. If your DAS requires external power (it is not powered by the USB connection), you should also put it on a UPS as shown next.

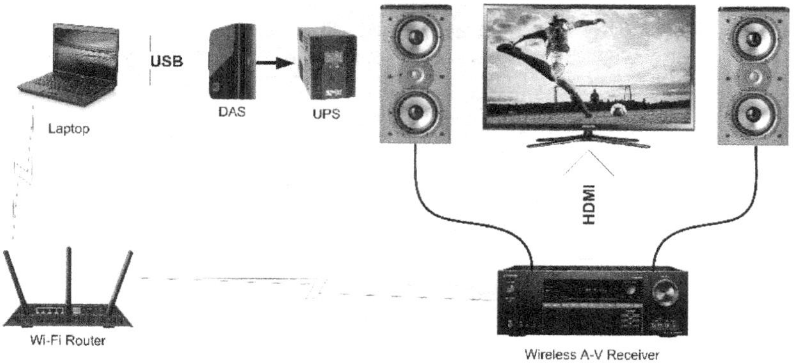

Figure 63 - Wireless Laptop Connections with DAS

If you decide to use a desktop computer instead of a laptop, the specifications and connections are similar but the numbers of pieces involved change dramatically.

Desktop Specifications

Desktop computers are larger and more powerful for the same price point of a similar laptop. Desktop computers require more physical space, generally noisier and hotter, and you must also purchase a monitor, keyboard, mouse for setup and administration, and backup power supply. However, you can stuff really big and multiple drives inside.

Item	Specification
Wi-Fi Card	Dual-band
NIC	10/100/1000
Wi-Fi Router	Tri-band
Processor	2 GHz or more
Number of Processors	2 or more
RAM	8 Gb or more
Internal Hard Drive	4 Tb or more

Figure 64 - Desktop Minimum Specification

While a minimum 4TB drive is recommended, remember that high-resolution audio/video files are very large and you may quickly run out of room. Adding a second hard drive to a desktop computer is usually a simple thing to do.

The supporting hardware components that you need for this approach are:

Component	Specification
Wi-Fi Router	Tri-band
Monitor	Yes
Keyboard	Yes
Mouse	Yes
UPS	Yes

Figure 65 - Desktop Supporting Components

For desktop computers, there is no need to buy a high-resolution screen, touch screen, or separate video card since this computer will be dedicated as your media server. Below is a summary in picture form of the required connections.

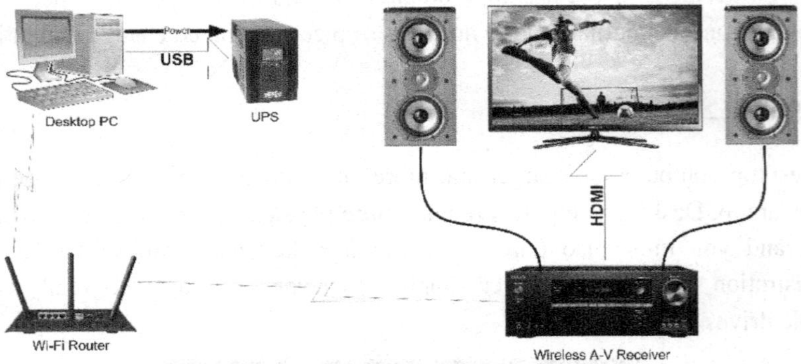

Figure 66 - Wireless Desktop Connections

If your desktop computer does not permit adding more or bigger internal hard drives you can also add a DAS as you did with the laptop through an available port. Again, if your DAS requires external power (it is not powered by the port connection), you should also put it on your UPS.

This is all fine and well if you have a wireless A-V receiver, but what if you don't? Do you need to buy one? Well, you could to make things easier but read on about the advantages and disadvantages involved in using hard-wired connections.

Using a Non-Wireless Receiver

This is where complications begin and why most folks shy away from other than simple wireless connections. But remember the upsides of hard-wired connections are that they are less noisy and capable of a higher data transfer rates. So let's see what the wiring differences are when using a non-wireless A-V receiver.

The easiest way to do this is to connect the output of the laptop or desktop PC through another HDMI cable to your A-V receiver. This gives you the best possible audio and video signals at the expense of having the laptop or desktop nearby your non-wireless A-V receiver.

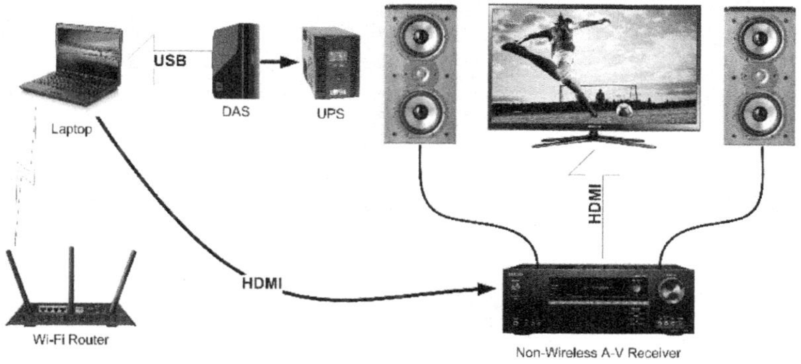

Figure 67 - Hard-Wired Connections

Another possible configuration uses individual A-V and audio connections to the receiver.

The Mini-ITX

So far, we have talked about conventional-sized laptops and computers. Another option for a small footprint computer is the Mini-ITX system. Several manufacturers produce these computers in various configurations but their draw is their compact size. Some are smaller than a Mac Mini but all are smaller than the smallest tower, Mini-ITX PCs are inexpensive, powerful, energy smart, and full featured however they do require a monitor and keyboard for setup and administration.

Much like the Mac Mini, the largest draws for a Mini-ITX are its low cost, small footprint, fan-less construction and built-in dual-band wireless network card. Capable streaming 4K video; DAS/NAS connectivity, and running a variety of OS, this may be the best choice for a media PC next to a laptop.

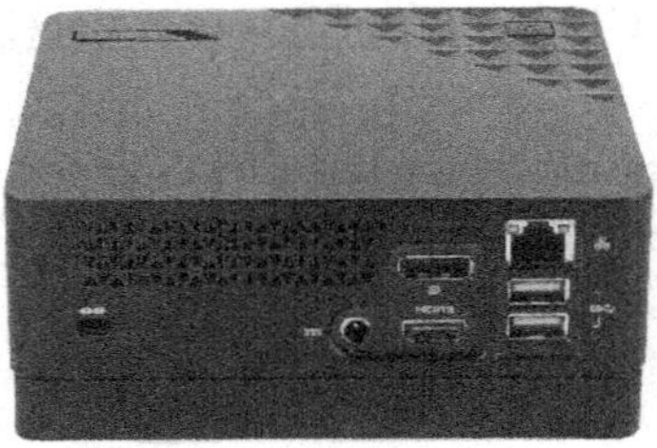

Figure 68 - A Mini-ITX PC

The downside to using a Mini-ITX computer is that its audio quality is like a laptop (less than optimal since the built-in sound card is the only audio output). However, they can also run a DAC making it an ideal computer for an upgrade path.

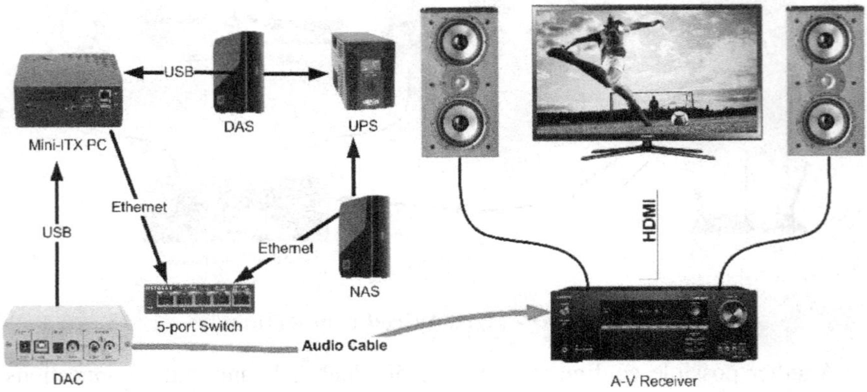

Figure 69 - Mini-ITX Connectivity

The above diagram shows the Mini-ITX configuration with both a DAS and an NAS connected. Also connected is a USB DAC providing the highest quality audio to the A-V Receiver. To make this a full streaming media system, all that is required is an HDMI cable connected between the Mini-ITX and the A-V Receiver. As with a laptop, it is also possible to use all-wireless connections as shown in Figure 54 or Figure 57.

Casting (Mirroring)

Casting (*screen mirroring*) is a slick feature for projecting video from a wireless device such as your cell phone or tablet to a compatible television or A-V receiver. Casting may require an additional piece of hardware and app to do this for devices lacking this feature and is covered in more detail earlier in this guide (see Google™ Chromecast™).

The weakest-link-in-the-chain regarding casting/mirroring is again that of matching the anticipated quality to the throughput speed of your Wi-Fi network. The recommended *minimum* throughput specs for Wi-Fi casting/mirroring video are 3.0 Mbps for SD, 5.0 Mbps for HD, and 25 Mbps for UHD quality.

There are other types of less complicated media servers that do not involve using so many components and that is to buy one ready-made (plug-and-play) as we shall see in the next chapter.

Plug-and-Play Media Servers

With an already substantial number of folks taking the plunge into media servers (you being another one of them), it is no wonder that some manufacturers are gearing up and producing plug-and-play (P&P) systems. P&P media servers come in all price ranges and configurations and combine everything you need into one box. However as of this writing the only high-end P&P media servers are capable of playing high-resolution audio files[2].

Figure 70 - A P&P High-End Audio Media Server

The upside of buying a P&P media server is that cabling is a piece of cake as is their typically smaller footprints. The downside is that P&P servers usually must be located near your receiver and they are not easily upgraded. Much like buying a car and then realize the entertainment system is integrated with the air conditioner, you must buy a new car to make an upgrade to the factory unit 9or return it to the factory for an upgrade).

And if you can bend the literal definition of a P&P media server to require the addition of an external storage device, there are a number of options available.

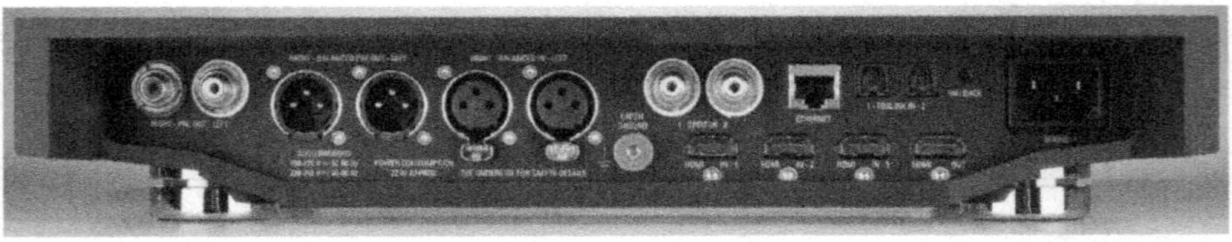

Figure 71 - Most High-End P&P Media Servers Require a DAS or NAS

---

[2] A few NAS media servers are the exception.

As mentioned earlier, when you think about all of the individual components parts put inside one box of a P&P Media Server (and one power supply feeding everything) you can see that flexibility is not designed into this type of solution.

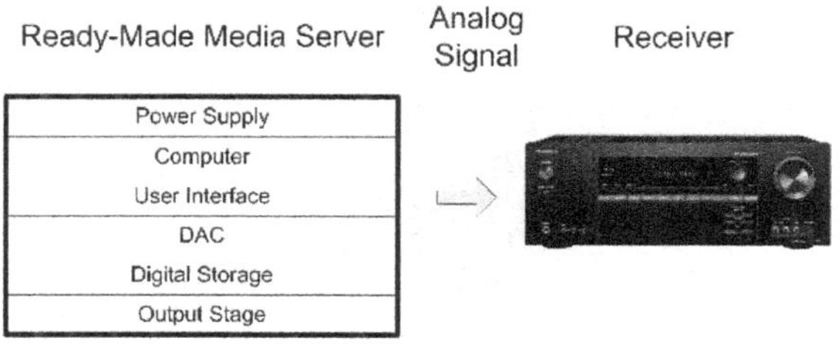

Figure 72 - P&P Media Servers Are Convenient but Inflexible

To get high-quality signals from such a device, you must spend a lot of money. Fortunately, there are some P&P systems that are exceptionally good but are also proportionally expensive. However, there is a way to do audio and video streaming from a component that may already be attached to your home theater: your *video player* or your *set top box* (*STB* as described in the next chapter).

There is an upside and downside for using your video player or STB as a media server but it is entirely possible to do this. The upside is that you do not need a laptop or desktop computer but the downside is that popular management software may not run on such devices. For example, the OPPO Digital series of video players (BDP-10x and BDP-20x) have the upside ability to stream both audio (in many high-resolution formats including DSD) and video up to 4K resolution from attached DAS devices (not truly an all-in-one solution but an "almost" P&P solution).

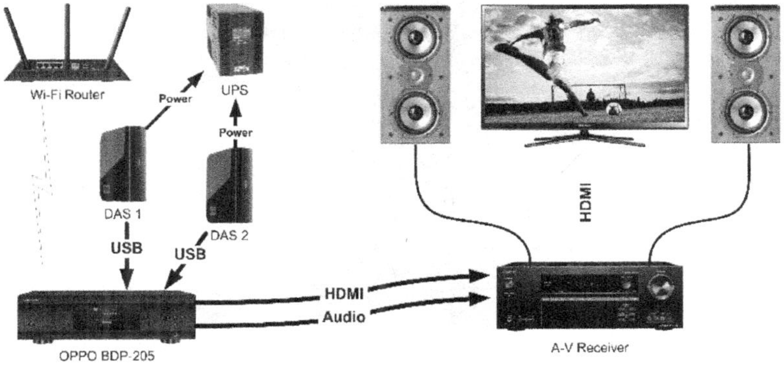

Figure 73 - OPPO Media Server Connections

Another upside of this approach is their small footprints and built-in apps for streaming internet media. Sound like a winner to you? It did for me and this is what I personally use giving the compromises I needed to make in my small home. One last upside of this particular unit is that is can stream from a NAS and your digital storage device can be located essentially anywhere.

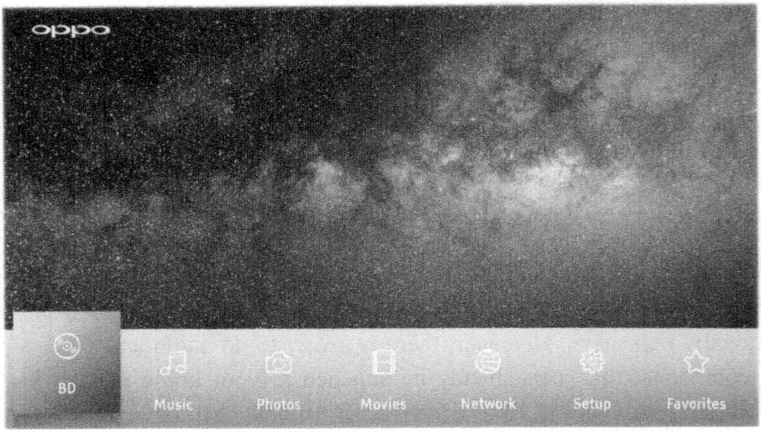

Figure 74 - OPPO Input Selection Display

The downside to this approach is that the user interface allowing you to control the OPPO is not as robust or user friendly as other dedicated media server applications. Rather, it browses the contents of your DAS just as Windows Explorer browses files on your computer. This supplication will be discussed in more detail later in this guide.

Figure 75 - OPPO Media Center App

The NAS Alternative

Some manufacturers have embraced the field of streaming media and created not only a network access server (NAS) but include enough intelligence within the NAS to completely function as a P&P media server. Some even offer solid-state drive caches for queuing media and improving fidelity and even HDMI and audio outputs (yes, some even have a built-in DAC) for direct connection to your audio-video system.

The upside of such a device is that it is highly reliable, integrates with software management applications (such as PLEX), and – depending upon its configuration – can store a huge amount of data. The downside is that it uses a similar audio output configuration to a laptop or an unbalanced audio cable and they can have somewhat noisy fans.

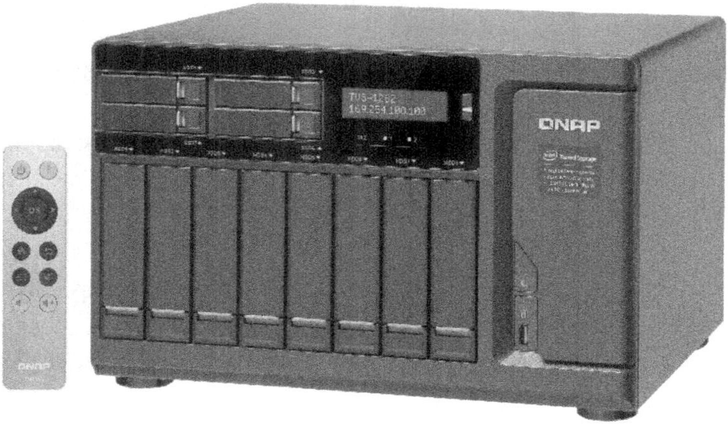

Figure 76 - A Large NAS P&P Media Server

Depending upon the NAS you buy, co-resident applications can automatically update your video library. For example, an application called Sickbeard (https://sickbeard.com) can be installed as a background service on QNAP, Synology, Netgear NAS devices that monitor network broadcasts for new episodes of your favorite shows. When a new episode is found, it downloads these shows into your library automatically.

Larger QNAP devices use a variety of free downloadable add-ons to enhance the function of their NAS. For example, their Media Streaming add-on installs into their basic QTS operating system that allows you to stream your media to different devices in different locations simultaneously using AirPlay®, DLNA, Chromecast™, and HDMI-connected devices. With the advanced management of the DLNA Media Server, you can also set advanced settings, such as DLNA client control, menu language and more.

Some NAS devices also offer multi-zone media access. With this feature you can stream all of your content on any streaming-capable device. Although the resulting audio and video quality is dependent upon the capability of that streaming device, the benefits are huge. Imagine watching a movie over HDMI, streaming a video to Chromecast or Apple TV, and even playing music on a Bluetooth connected speaker, all at the same time all from the same NAS[3]. Management of this NAS is also possible through some of these same streaming-capable devices thereby eliminating the need for a separate administration computer.

If you do not have a NAS, another application called another application called Sonarr (https://sonarr.tv) can be installed on Linux, Windows, and Mac computers that will do pretty much the same thing. Support for both of these applications is volunteer-based so if you have issues installing, using, or integrating them in your media server you cannot speak to a live support person.

Another style of less expensive P&P media servers can be grouped into the "pay for services" types called *set top boxes*. Here there are many from companies like Apple, Google, Roku, and NVIDIA all vying for your attention as we shall see next.

---

[3] As mentioned earlier in the chapter on Networking, remember that simultaneous multi-user streaming can overload your router's throughput and cause dropouts or delays.

Internet and OTA Streaming

Internet Streaming devices are the last gizmo add-ons to your streaming media system I wanted to bring up because of their popularity and convenience, especially when streaming video content with newer Ultra-high definition resolution (*4K*) TVs. Some of you may already stream video from your current Smart TV and internet connection and have thought nothing special of it (Netflix, YouTube, etc.).

These built-in features are called streaming apps but there are a host of manufacturers out there who offer add-on set-top boxes (STBs) for you if your system does not offer a particular app (or does not have any apps at all). STBs are very compact in size and stream through the HDMI cable however some contain a DAC that can also stream audio through a headphone jack (see Figure 38).

Set Top Boxes

Focusing more on video than audio, these devices can only stream a limited number of high-resolution audio files (few play FLAC and none play high-resolution FLAC or DSD) but most offer 4K video streaming. With that said, this one of the fastest growing niches in streaming media and this is because STBs offer some fantastic advantages over other approaches. But there are disadvantages too.

Each STB will stream a host of video from the internet (*IPTV*) providing thousands of channels from which to choose (the upside) but each of these channels is a subscription (downside). Some subscription services like Netflix offer commercial-free video streaming (upside) and some do not. Here are a few of the choices you have when considering a STB purchase to complement your streaming media system.

Subscription audio and video streaming services are all compressed in some format, although a few have surfaced (Tidal, Deezer) to stream uncompressed audio.

Google™ Chromecast™

Chromecast is a great all-around option that will work for anyone with a Smartphone. There's no remote in the box and no on-screen menu, so everything is controlled from the app, which is widely available and free to download. The regular

Chromecast is one of the cheapest ways to get media to your HDTV, while the Chromecast Ultra will deliver high-end content to match your brand new 4K UHDTV.

Figure 77 - Google Chromecast Set Top Box

The Chromecast plugs into an HDMI input on the back of your TV and the internet connection is on the power supply (a nice feature for those who hate looking at all those cords).

Amazon™ Fire™

Figure 78 - Amazon Fire Stick Internet Streamer

For anyone invested in Amazon's Prime content (Instant Video, Prime Music), Amazon's Fire is a good way to go. You also get the use of Amazon's Alexa and Apple's Siri voice-controlled assistant, whether you go with the Fire TV (4k video, high performance gaming) or the more affordable Fire TV Stick, thanks to the included remote with voice support.

On the other hand, Amazon's Fire TV ecosystem is based on the Android OS. This does not imply that every Google Play app you run on your Android Smartphone will also run on the Fire. That said, you *do* get BBC iPlayer, ITV Player, STV Player and Demand 5 - no Channel 4 catch-up apps for now; this too can and undoubtedly will change in the future.

Figure 79 - Amazon Fire Set Top Box

On the plus side, the Fire STB also plays 24/48 FLAC (no dither) from a DAS but limits audio output to the HDMI cable.

Apple TV

If you've got an iPhone, an iPad, a Macbook—the Apple TV is a no brainer for your streaming needs. Apple has always been great at integrating their products, and the AppleTV fits in nicely. You get similar streaming options to other devices, but you'll also be able to share the screen from your devices on your TV using the Airplay option, play games, and control everything with your voice thanks to the Siri voice control built into the remote.

Figure 80 - Apple TV Set Top Box

The current models come in 32 Gb or 64 Gb versions. Both include the same features, the latter offering more internal storage if you want to download more apps and games.

Roku

If you're looking for something to match well with your 4K TV, the Roku may be what you've been after. The Ultra model offers 4K HDR visuals, and more streaming options than any other device. It can even help you find your remote if it goes missing. Their Express+ add-on offers non-HDMI (*composite*) AV connections for even the oldest of TVs to economically turn it into a smart TV.

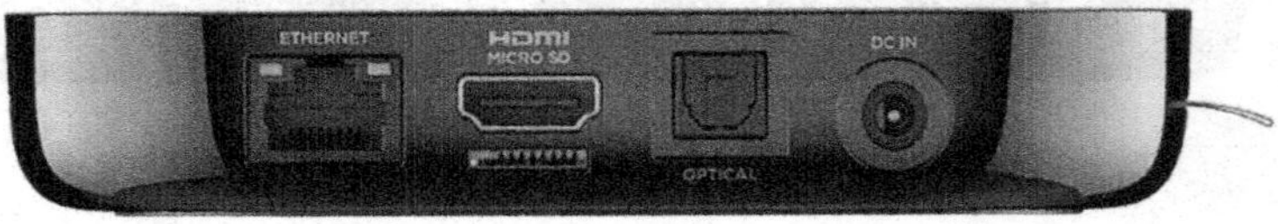

Figure 81 - Roku Ultra Set Top Box

Roku STBs have an edge on their competition by permitting access to a NAS via the "Q-media" channel via the DLNA protocol. The QNAP folks, makers of fine NAS devices, provide instructions on how to do this here: https://www.qnap.com/en-us/how-to/tutorial/article/how-to-play-media-files-on-a-qnap-nas-with-roku-channel-qmedia.

By adding the Roku "Media Player" channel, it can also find, browse, and play content from media servers on your local network as well as USB (on those Roku players with a USB port). It supports such media servers as Twonky, Windows Media Player, Plex, and Tversity.

The Roku also is one of the few STBs that offer an optical audio output (TOSLINK) for a high-quality audio connection to your A-V receiver.

Android

For you Android fans, not to worry; you are not over-shadowed by the big-name manufacturers in the industry. Unlike one manufacturer of one product, there are a host of Android devices from a wide variety of manufacturers from which to choose.

Not only are the cost of these devices extremely low (the operating system is free), they also offer features other STBs do not. For example, the Comigo STB also provides audio output via an internal DAC and RCA jacks.

Figure 82 - An Android-Based Set Top Box

OTA Streaming

This is a category of which many of you may not be familiar called over-the-air (OTA) broadcasting. Historically, this is how television became popular and it requires a small piece of hardware (much like adding a STB) called an antenna for its use. As it was with STBs, there are upsides and downsides of OTA devices.

The biggest upside is that once you buy the antenna, there are no additional fees. That's right, OTA allows you to completely "cut the cord" and yes, the service is completely free, free, free! The downside of OTA streaming is that you do not have access to specialized subscription programming like the Disney Channel, the History Channel, HBO, etc. And, by the way, one final upside is that OTA Streaming is *uncompressed* video unlike other satellite and cable media streaming devices.

Antennas

The only thing you need to watch OTA streaming is an antenna. Here as with STBs you have many choices all of which depend on your location and your type of home. Antennas require what is called *line-of-sight* aiming to the transmitting tower; the best picture is achieved by pointing the antenna at that tower. In rural areas (50-100 miles from the tower), high-gain directional antennas and booster amplifiers may be required. These types of antennas are physically large and must be mounted outdoors on long poles. A single cable connects an antenna to the back of your TV.

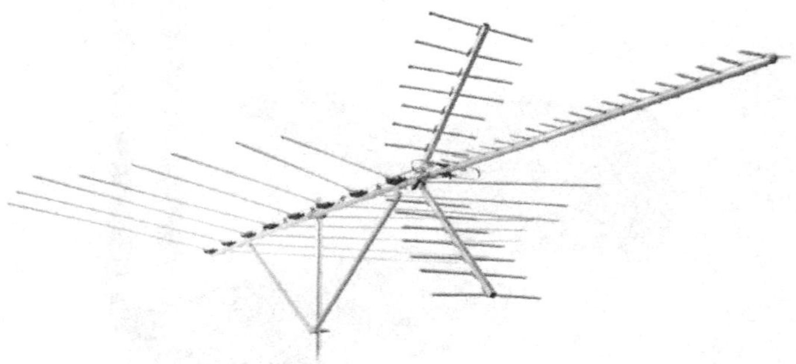

Figure 83 - A High-Gain Rural Directional Antenna

But in metropolitan (non-rural) areas closer to the transmitting tower an omni-directional antenna will work just fine. These small antennas can hang on a wall or even laid flat on a table top. The only issues with such small antennae are that some are susceptible to signal loss when used inside of metal or concrete buildings.

Figure 84 - An Amplified OTA Omni-Directional Antenna

Both approaches offer amazing high-definition signals that are uncompressed and not subject to the usual interruptions experienced with internet or satellite streaming services such as power outages and of course cloud cover. OTA streaming is especially useful during severe storms where no other form of service may be available.

OTA streaming offers real-time viewing for free and yes, there are products that will allow you to record broadcasts and play them back when you have the time to watch them. These devices are called OTA DVRs.

OTA DVRs

Most people today have very busy lives and cannot be present at the time a live broadcast occurs. Digital Video Recorders (DVRs) allow you to time-delay viewing so you will not miss a televised program. Much like an internet STB, these devices connect to the antenna and then another cable (HDMI or coaxial) connects to the inputs on your TV. As of this writing, there is only one device that is totally subscription free available and that is made by ChannelMaster.

ChannelMaster

ChannelMaster has been around for decades offering OTA antennas for use since television was first broadcast. These folks really understand what it takes to capture signals and do it right. Taking this understanding one step further, they now offer a DVR capable of doing more than just recording free broadcasts.

With such a device you can watch and record free, live OTA broadcast programming in crystal clear uncompressed HD, including all of your favorite shows, news and sports from ABC, CBS, FOX, NBC, PBS, The CW and many more. ChannelMaster DVR also offers familiar menus and a program guides similar to those used by cable and satellite services, and dual tuners allowing you to record one program while watching another or record two programs at the same time.

Figure 85 - An OTA DRV

If you are considering buying a STB, you may think again. This DRV also offers Internet connectivity to provide additional features including an expanded 14-day program guide, automatic software updates, access to online streaming services and a free subscription to Channel Master TV (30+ Internet Channels) and a high-quality optical audio (TOSLINK) output jack.

The feature I personally value most from any STB or OTA DVR is a high-quality audio output. Such devices require additional connections and we shall see discuss the differences in audio, video, and AC power connections in the next chapter.

Connections

We've discussed a lot about hardware in the previous chapters and shown some diagrams on how all of these things cable together. In this chapter we will discuss the details of high-end cabling and other electrical-related issues.

Audiophiles have found that swapping the interconnecting cables between different pieces of gear can change the way things subjectively sound and visually appear. Most engineering types, however, strongly disagree and typically respond by stating something like "that cannot be; wire is wire." These folks usually point to the "double blind" test as the preferred scientific method for proving or disproving subjective claims. What derogatory responses of this nature are *really* saying is this: "Can the human eye/ear consistently detect differences?"Although there is some science behind the construction of exotic cables, visual and audible results remain a hotly debated item.

Even though small differences can be measured, the proof is in the ability of the human eye and/or ear to detect them. For example, I took a high-resolution photograph (on a tripod in the middle of the night using identical exposures) of the same high-quality video image played through my OPPO BDP-105 Blu-ray player: one photograph of the screen from HDMI cable A and one from cable B. I then used a software application (Paint Shop Pro) to subtract the common information between the two photographs. The un-retouched un-enhanced results are shown in the far right-hand image below so the differences are not emphasized in any way.

Figure 86 - Photographic Difference of Two HDMI Cables

Can you see the difference between the two photographs without looking at the difference image? The camera was able to easily detect small changes between these two images and the software program shows the color black in areas where no change was measured. Although only the best eyes would be able to identify these subtle changes, one wonders if even trained high-end eyes could do so in a double-blind test. So the double-blind test has merit – up to a point.

The following image is an enhanced version of the difference image (inverted colors of white-to-black and black-to-white) so that you can more easily see all of the differences the camera measured.

Figure 87 - Larger Processed View of the Difference Image

Now knowing where to look for these differences, go back and see if you can observe them. For example, look at the upper-right-hand corner of the original unprocessed difference image, you can easily see a speckling of information around the rocks, the upper-left-hand corner shows a color change on the "pause" button, and the lower center shows a color change of a "gold line."

This proves without a doubt that there are changes in a video image when changing only an HDMI cable. The average person is untrained to notice how and where these differences exist. Does this make disprove subjective evaluations? In my opinion, this proves the difference between a discerning eye and ear (subjectivity is a matter of training). I'm sure that similar things could be measured when swapping audio cables and using precision microphones and test equipment.

I personally find great value in auditioning cables at every opportunity to see what impact they have as my system evolves. I ignore all manufacturer's claims, sales hype, and prices, and simply observe the differences hopefully without bias or expectation. Every time a new piece of gear makes its way into my system, I usually notice that something changes – typically something very subtle – either audibly or visually or both. Much like putting a very large high-performance engine in a small car, your eyes and ears can notice changes when they are big enough. But the best changes you will notice are when the weaker links in your A-V chain is revealed.

As your system evolves and its quality improves, what you noticed yesterday may not be what you will notice today. With enough time, you train your eyes and ears to become keenly aware of minor differences as your knowledge of these subtleties improves. You become what is called a "golden ear" or videophile. The hardest thing to do is keep biases at bay so your mind is open to observe these more subtle differences. With this said, let's look at some of the options you have when connecting cables to your system.

Interconnects

Although others exist, there are five popular styles in use today for connecting audio/video from one component to another:

1. RCA cables (*unbalanced* audio cables)
2. XLR cables (*balanced* audio cables)
3. S/PDIF cables (*optical audio* or *TOSLINK* cables)
4. HDMI (high bandwidth audio/video cables)
5. Digital Serial (USB, Thunderbolt, Firewire, etc.)

It is unnecessary to spend time on the permutations, winding techniques, shielding, and conductive materials used in each of these cables with the exception of mentioning that each *can* matter, especially shielding. The other physical characteristic that matters is the length of the cables. Here, longer cables exhibit more losses, can potentially color the sound, and potentially distort the picture than shorter cables. So in brief, shorter is better.

But there is a compromise you must make when thinking that the shortest cable is the best and that is planning for future changes. If today your media server is located ½-meter from your A-V receiver, this may change in the future and you may wish to consider buying a 1-meter cable instead.

The other issue is in necessary long cable runs. For some reason, your DAS may need to be connected with a very long USB cable.

Similar issues present themselves in other connections to your system.

Power Cabling

For the longest time I was a personal disbeliever that power cable changes could somehow make a difference in quality. After all, wire is wire and there is only one frequency (50 Hz or 60 Hz) that must be considered, right? Wrong! First of all, this assumes that there is only one frequency coming out of the wall outlet in your home and this is woefully incorrect. In truth, electrical noise is everywhere and depending on the location of your home your entertainment center can have "dirtier" power than the home across the street or even the room at the opposite end of the house.

As mentioned earlier, the other place noise can enter your electrical power is through the wiring in the walls of your home. To understand this issue better, try to imagine of all of the wiring hidden in the walls in your home. Every length of wire acts like an antenna and airborne noise like cell phones, radio and television stations, etc. into your electrical power. This noise when improperly eliminated can make its way into the electronics of your system and literally wreak havoc.

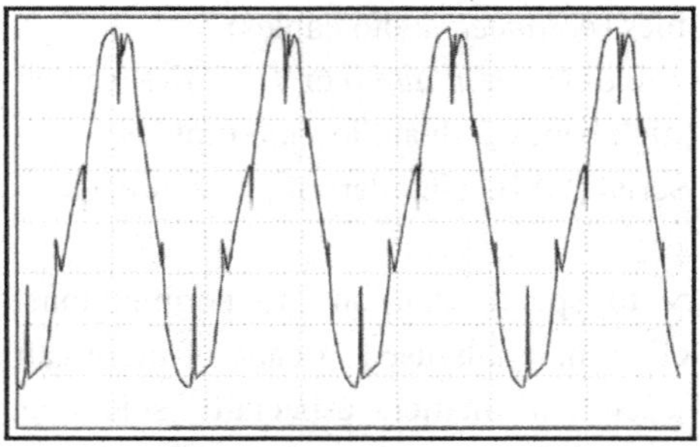

Figure 88 - Electrical Noise on 60Hz Power

I admit that I had a closed mind and dismissed this as snake oil. What I did not consider is all potential places I can look to eliminate this noise, starting at the wall outlet. The power cable between the wall outlet and a piece of gear is the first place under my control where I can help eliminate this airborne noise problem.

Tracing back where the wall outlet gets its power is the wiring in the wall and this is what it is; I have very little or no control over that. Short of hiring an electrician to run a dedicated and shielded power line from your circuit breaker panel to your system, you are at the mercy of this in-wall wiring and the signals in the air.

Your system will work just fine using conventional power cords but as it was with changing HDMI cables, changing power cords can have a similar effect. Again this remedy can strengthen a weak link in the chain.

But speaking of the house wiring, there is one more issue that you *do* have under your control that you can control: the electrical line phase at your wall outlets.

Electrical Line Phases

For some countries, this is not an issue since wiring in them is all 240V and all wall outlets are wired in the same electrical phase. But in other countries like the USA, electrical power in homes use two 120V lines (L1 and L2) wired 180° out of phase with a neutral reference point. 120V wall outlets in homes are wired to either the L1 or L2 phase to balance out the load on each phase at the circuit breaker box.

Normally, this is not a problem but when eliminating noise from all sources, powering some of your equipment from L1 and others from L2 generates more noise than if all of your equipment was connected to the same electrical phase It really doesn't matter which phase you use, just that all equipment is on only one phase.

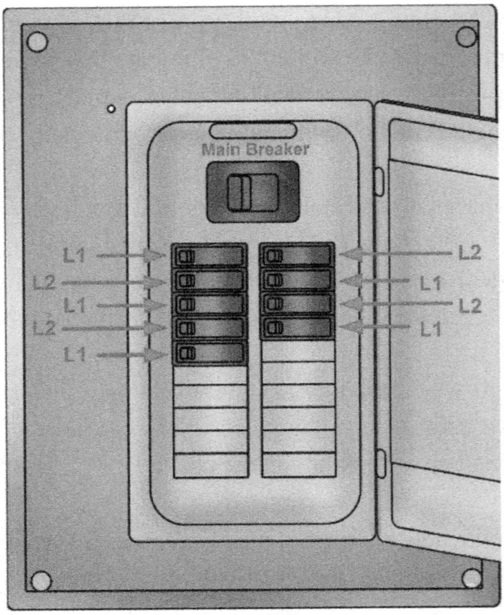

Figure 89 - L1 and L2 Phase Distribution on a Circuit Breaker Panel

The way to test this is to turn off your circuit breakers one at a time and find out which ones turn off the wall outlets to which your system's components are plugged. Notice that the **column** of circuit breakers toggles back and forth between L1 and L2. Also notice that every **row** of circuit breakers is either L1-L2 or L2-L1. If the results of your test shows one outlet is on phase L1 and another is on phase L2, have an electrician move that circuit breaker to the other phase so all are on the same electrical phase (which phase does not matter).

Once cabling and electrical issues are resolved, you can be reasonably certain that any noise in the system is most likely coming from the internal electronics. As mentioned earlier, there are a few manufacturers that may provide power supply upgrades for your equipment. You may also consider trying a power supply.

Finally it's time to install the management software, download some digital media, and preserve you're existing A-V library in a high-resolution format.

Software

Congratulations! You now have a media server worthy of streaming high-quality audio and video. Attention to detail on the front-end of your setup determines how well your system can perform. Finding the right cable combination and eliminating noise from AC power can improve the sound of your system. Take a few moments now to double check your work.

TIP:

I have a habit of making lists. If your system is presently at an intermediate state (starting with this but eventually want it to be that), I recommend writing down a *migration plan* showing the transition from where it is now to where you want it to be (replace cables first, then add a Line Conditioner, upgrade the DAC, etc.). Do this in the order you suspect you will make your upgrades but remember that this order will undoubtedly change as your sensitivity to its subtleties improves. I also have a small savings fund to which I contribute a little extra money each month for the exclusive purpose of buying this upgrade equipment. Then I can draw on this fund when I run into a deal that I "just cannot refuse."

Most likely you want to listen to the fruits of your efforts as soon as possible. You can test your system by using the built-in operating system apps (Windows Media Center, Apple iTunes, etc.) to play your existing music files. However, these apps most likely will not play high-resolution audio or video files (i.e., they can only play compressed sources such as MP3, AIF, MP4, etc.). What you now need to do is to abandon those old apps and compressed files and install new software that will manage and play your entire high-resolution (and low-resolution) library.

By now you should also have narrowed down what your target audio format will be (FLAC, DSD, etc.) and you should have all of your hardware setup and in place. Only a few things remain to make your media server completely operational:

- Pick the management software you intend to run on your media server
- Rip your existing CDs, DVDs, and Blu-rays to a lossless format
- Start buying high-resolution downloads

Initial Configuration

The first thing you must do before you start installing any management software is to decide **where you want to put your audio and video libraries.** I understand that you want to put your libraries on either your built-in drive or DAS or NAS, but that is not what I mean. You need to decide the name of the folder (the permanent "HOME") on that storage device where these files are to be stored. No matter what software application you use to manage your libraries, this is where on your storage device HOME will be.

For example, upon installation the iTunes software creates a HOME on your computer's hard drive as: /Music/iTunes/iTunes Media/Music/. If you use the Macintosh Finder you will see that iTunes adds all new downloaded music under that HOME folder as: */Artist/Album/Track_Song*. The path to any individual song on a storage device is then:

/Music/iTunes/iTunes Media/Music/*Artist/Album/Track_Song*

Figure 90 - iTunes HOME Structure

Whatever management software you choose (as described below) it will ask you this same question during its installation process (something like "Where do you want your music to be stored?").

I personally use a less-complex, one-word folder structure on my storage device than the iTunes default HOME suggests. I also separate the type of information I'm storing by those names as shown next.

- **/Music/***Artist/Album/Track_Song*
- **/Video/***Title*
- **/Pics/***Event/Title*

Figure 91 - My Personal HOME Structure

All you have to do when adding media to your library is to remember to put all of your files under this HOME (e.g., downloaded high-resolution audio files or ripped CDs and Blu-rays to this location). Speaking of your management software now is the time to make your choice of which one to use.

Management Software

The last choice you must make is this: what software application you want to use to manage your new digital library? This software is the interface you will use to access the media and will probably be a tougher choice than all of those you have already made. For this reason, I have prepared a list of some of the more popular software applications highlighting the Pros and Cons of each. This should give you an idea of what to look for when making this final choice.

Another problem you may encounter is that once you realize how much content you have, you may not be able to find the music or video you want to enjoy. My personal music collection consists of over 200 artists each with between two and 20 albums and each album with 12-36 songs (that's about 18,000 songs…and growing). Finding a particular song I wish to hear relies on my memory of its title or on the name of the album on which it appears. For this reason, searches and playlists are two features I would highly recommend in whatever management software you pick.

Also realize that starting with one application and switching over to another is not that big of a deal (that is, if you change your mind, all is not lost and you do not have to start over). Although the intricacies of applications differ, the process of making the change is usually a reasonable and uncomplicated thing to do; re-learning the operational differences, however, is.

Unless otherwise specified, the A-V file formats supported by all of the products described in this chapter (to the best of my knowledge) are:

Audio Formats

- AAC/HE-AACv2
- AIFF
- Apple Lossless
- FLAC (up to 24/192)
- HD-AAC
- MP3
- Ogg Vorbis
- WAV
- WMA

Video Formats

- 3GP
- AVI
- Blu-ray
- DVD
- GIF
- JPG
- MKV
- PNG
- TIF

JRiver Media Center

I belong to an audio club and JRiver is the most popular management software by far among those members streaming digital audio/video. This software has a very nice user interface, can run on pretty much any platform you desire, and can be wirelessly controlled by tablets and cell phones.

Website	https://www.jriver.com/
OS	Mac, Windows, Linux
Formats	APE, WavPack
Cost	Free to try $50 US - $70 US to buy
Pros	• Bit Perfect output • True 64-bit processing • Caches files to RAM
Cons	• Upgrades are not free
Comments	Higher-priced version permits multiple installations and on multiple OS

PowerDVD

PowerDVD is an outstanding choice for the Windows OS since it plays a wide variety of video, audio, and photo formats, includes numerous features, and has top-of-the-line performance on its extremely intuitive interface.

Website	https://www.cyberlink.com
OS	Windows
Formats	4K streaming
Cost	$265[4]
Pros	• Bit-Perfect Output
Cons	• No media store • Casting doesn't work with protected media • Doesn't work with Netflix • Upgrades not included
Comments	MSRP is high compared to others

WinDVD

WinDVD allows you to watch a large movie collection of both physical and digital formats all in one place. It lets you optimize video and audio settings and is packed with features so you get an excellent movie-watching experience.

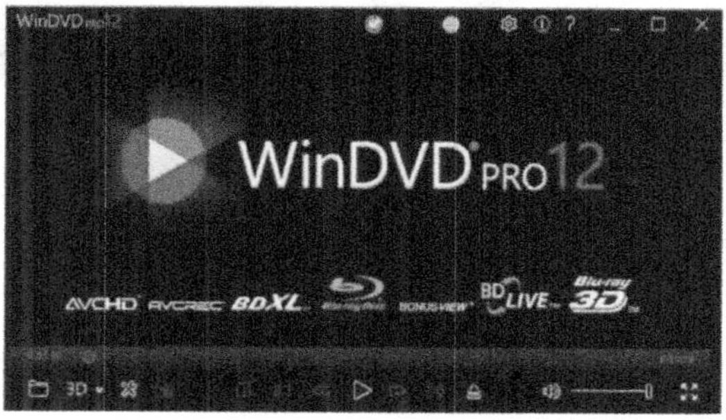

[4] Often discounted to $70

Website	https://www.windvdpro.com
OS	Windows
Formats	4K streaming
Cost	$80[5]
Pros	• Online movie search • 240 frames/sec.
Cons	• Upgrades not included • Only runs on Windows
Comments	Excellent video streamer but current version does not manage audio

Audirvana +

Audirvana plus is a well-designed Mac-only software that offers handy features like automatic sample rate switching and memory play. Supporting high-quality DSD and FLAC audio playback, it ranks very high for an overall top performer. Most importantly it does most (if not all) of the background optimization for you and setup time is minimal. Just install and enjoy (options are available but the recommended defaults usually provide the best performance).

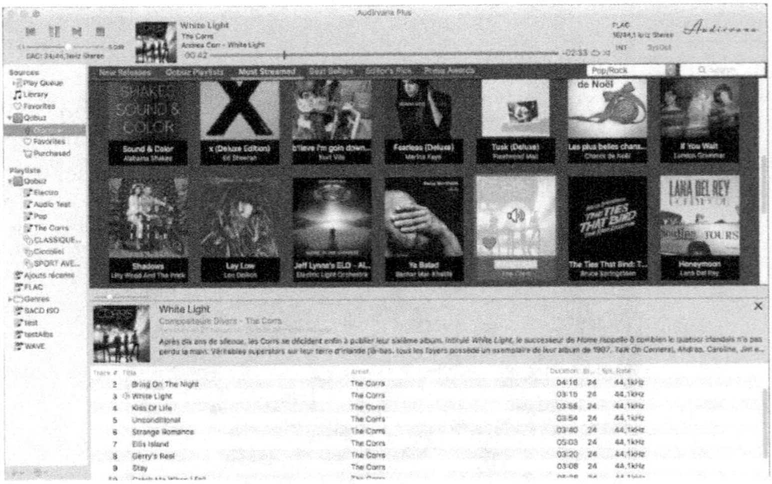

[5] Often discounted to $60

Website	https://audirvana.com/
OS	Mac
Formats	DSD, SACD
Cost	$74 US
Pros	• MQA audio decoder • Integrates with HQ streaming
Cons	• Only runs on Macs
Comments	No plans for supporting other OS

Plex

Do you own an Alexa and want to integrate it with your streaming software? Well here is an option for you not only for this unique feature but also for its ability to do much for so little. With its ability to store content in your Cloud or OneDrive, it offers another option that other programs currently do not.

For those of you who wish to "cut the cable" cord, Plex may be the way you want to go. Although pricing its features like many Set Top Boxes where you buy only those services you desire, Plex integrates your complete media experience in one convenient package.

Plex also easily integrates with many DAS and NAS devices making it flexible and expandable as your streaming needs change.

Website	https://www.plex.tv/
OS	Mac, Windows, Android, Linux, and many NAS
Formats	Only FLAC
Cost	Trial: Free, fee-based enhanced features Plex Pass: $5-$150
Pros	• Fast and intuitive user interface • Easy setup • Provides cover art, music videos, trailers
Cons	• Optional features at an extra cost • Does not support DSD audio • Requires high-bandwidth connection • Expect price increases
Comments	Strong app integration with Xbox, PlayStation, NVIDIA SHIELD, and Smart TVs

Universal Media Server

Universal Media Server is a DLNA-compliant UPnP Media Server originally based on the Play Station PS3 Media Server. This free application offers great stability and far-reaching file-compatibility.

As it is with any open-source software, upgrades and revisions are accomplished by volunteers. Support is forum-based but still pretty decent considering that it too is free. Replies to inquiries may be too complex for novices to understand and response times for involved questions can take days.

Website	http://www.universalmediaserver.com/
OS	Mac, Windows, Linux
Formats	
Cost	Free Monthly fee for Enhanced features
Pros	• Fast browsing • Highly customizable
Cons	• Confusing setup • Limited plug-ins • No remote access
Comments	Amazing performance although it currently does not support DSD

Subsonic

With a web-streaming (i.e., remote-access) design focus, Subsonic allows you to access your library from anywhere there is an internet connection. This is a very convenient feature for those of you who travel a lot. However, internet bandwidth limitations imply a resulting lower-resolution (i.e., lossy or compressed) video and audio from these remote locations.

Website	http://www.subsonic.org/
OS	Mac, Windows, Debian/Ubuntu, Linux, Synology
Formats	Podcasts
Cost	Free $1/month for enhanced services
Pros	• Supports large music libraries • Moderately easy setup
Cons	• Fee required for video support • No subtitle support or stream quality settings • Less robust support community
Comments	A good start for venturing into the streaming world

Serviio

Although capable of running on a streaming media PC as it is with all other management software, Serviio can also integrate into your NAS (that's right, you can eliminate the PC and you only need a specific kind of NAS).

Serviio is a very reasonably priced option for managing all sorts of audio and video formats, even those in high resolution. With its remote-access capability, you can view and listen to anything in your libraries from any internet connection. As it is with Subsonic, web-streaming your digital library from a remote point lowers its audio and video resolution but local streaming maintains high resolution.

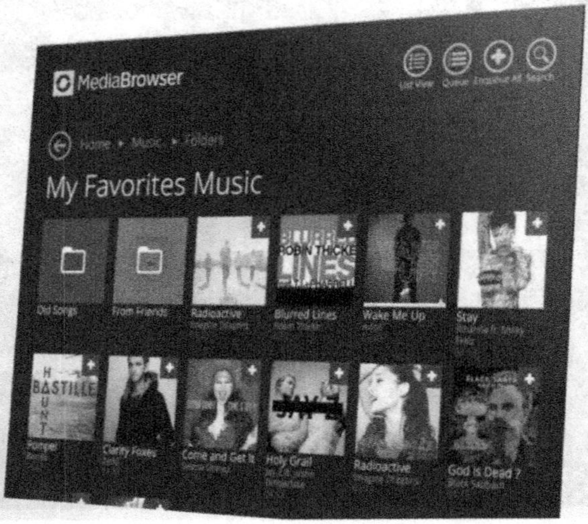

Website	http://serviio.org/
OS	Mac, Windows, Linux, Qnap, Synology, Asustor
Formats	DSD/DSF
Cost	$25 US
Pros	• Solid free/open source version • Remote access • Strong plug-in support • Integrates with trak.tv
Cons	• Top features restricted to Pro version • Confusing setup • No iOS app
Comments	Great remote access

<u>OPPO Media Control</u>

Not exactly in the same category but one worth mentioning is the OPPO Media Control application coupled with an OPPO DVD/Blu-ray/4K player. Since the OPPO media players are already a self-contained computers, the only things required for its operation are a user interface, here their Media Center app, and a digital storage device (DAS or NAS). In this regard this software functions much like Serviio does on a specialty NAS.

This is the approach that I am currently using. The OPPOs also contain high-quality built-in DACs and may also be a good option for you especially if you have a limited amount of space you can allocate to your streaming media system.

Website	https://www.oppodigital.com/
OS	Mac, Windows, Android
Formats	DSD[6], SACD 4K, Blu-ray, DVD, MKV, AVI
Cost	Free (this is an app only)
Pros	• No computer required • Little setup required • Supports very large music libraries • Easily connects to DAS or NAS devices • Fixed and variable volume output selection (option to replace your A-V receiver with a high-quality amplifier)
Cons	• Cannot control library metadata • Older models use 100 Mb LAN cards • Does not rip CDs automatically to your library • Gracenote lookup sometimes causes problems
Comments	This is a very frugal way to approach streaming

Sickbeard

Not a true media server software program but an add-on application, **Sickbeard** may be something you wish to install on your QNAP, Synology, Netgear NAS devices. With this application, your NAS can automatically update your video library for those televised episodes you frequently watch. For example, if you are a fan of the PBS series *Downton Abbey* or the SciFi series *Doctor Who*, this application monitors whatever program you tell it to (a watch list) and downloads it to your NAS automatically when a new episode is found.

[6] Earlier models only support DSD64

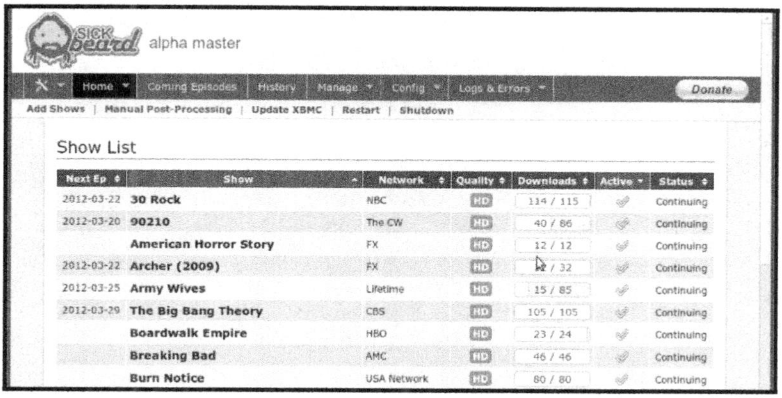

Website	https://sickbeard.com
OS	QNAP, Synology, Netgear NAS
Formats	
Cost	Free (this is an app only)
Pros	• No computer required • Little setup required • Automatically updates television episodes
Cons	• No live technical support • No option for live technical support
Comments	Requires a NAS

Sonarr

If you decide to install PLEX as your media server software, you may want to consider this useful add-on application. Similar in function to the Sickbeard application, **Sonarr** pretty much does the same thing. With this application, your computer can automatically update your video library for those televised episodes you frequently watch. For example, if you are a fan of the PBS series *Downton Abbey* or the SciFi series *Doctor Who*, this application monitors whatever program you tell it to (a watch list) and downloads it to your computer automatically when a new episode is found.

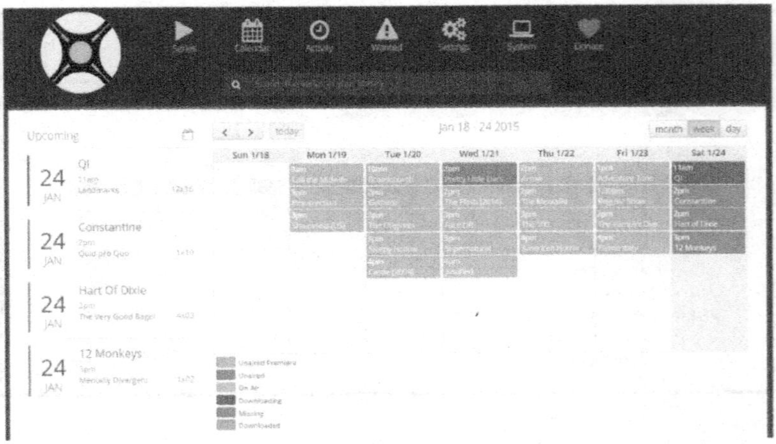

Website	https://sonarr.tv
OS	Linux, Windows, Mac
Formats	
Cost	Free (this is an app only)
Pros	• No computer required • Little setup required • Automatically updates television episodes
Cons	• No live technical support
Comments	Requires PLEX

What I did next was to replace all of my *lossy* audio and video by re-ripping my entire A-V collection with a better software application in a *lossless* format.

CD Ripping Software

Some software management applications also have built-in features to rip CDs (and sometimes even Blu-rays) into your digital library. A word of caution is due here in that the method used to rip content may not be of high quality. For this reason I suggest you choose to use another application to rip your media into your library.

For example, I had over 600 CDs that I had previously ripped to MP3 format. I still had all of those old CDs and after tossing out those I no longer wanted, it took me about two weeks to re-rip all of them to FLAC (free lossless audio compression) format. While there are many free applications available to do this in a *compressed* FLAC form, the **dBpoweramp Reference** (https://www.dbpoweramp.com/) software rips CDs to an *uncompressed* FLAC form. This creates larger files and dictates the need for more storage capacity but the results are very audible and well worth the time.

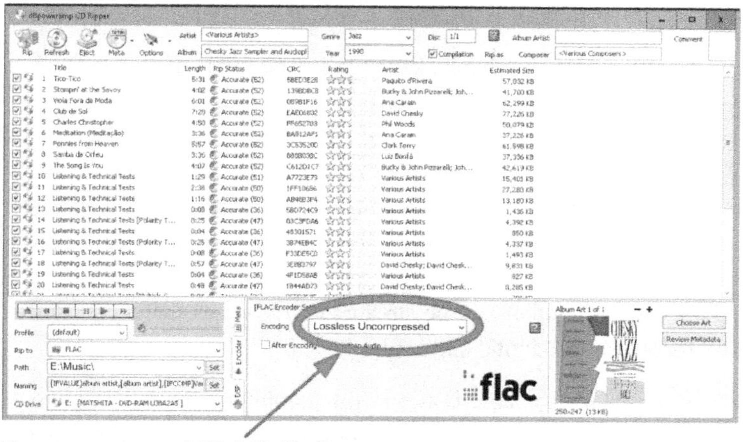

Uncompressed FLAC Option

Figure 92 - dBpoweramp CD Ripper

You can choose whatever software you wish to re-rip your audio library but be aware that not all software is created equal. I only wanted to do this twice (the first time AND the last time) and I imagine you would prefer to do the same.

After purchasing and installing your CD ripping software, you should point the output to the HOME location of your library (see Initial Configuration earlier in this chapter). This way whenever you rip a CD it will automatically put it in the right place. For example, the PATH option in Figure 83 is set to the library location of **E:\Music** meaning this is the location of your digital audio library.

Before ripping your entire library, I would rip one CD as a test to make sure that you are getting the ripped output (especially the naming convention) you desire. I also encourage you to compare the original CD and the ripped FLAC version just to see if you can detect an audible difference. You may be surprised by the results and decide to go with a different ripping software.

The next step is then to feed all of your CDs – one by one – into your computer's optical drive to build your high-resolution FLAC music library. Although better quality than your lossy MP3 format, these new music files are *not* the highest quality you seek. Once your system matures, you even may decide to replace these low-resolution files with a higher quality audio by purchasing this same music in high-resolution format from one of many download sites.

Video Ripping Software

Ripping your DVD/Blu-ray/4K library is something you should also consider doing. While there are a plethora of folks who make good audio ripping software, there are only a handful of folks that make good video ripping software. To rip video, you must use an optical drive connected to your computer capable of playing your video disc. Adding the right software to this computer then permits you to create digital video files that can be streamed by your media server software much like it does for streaming digital audio files.

Ripping video has the same issues in lossy compression that ripping audio does and an original-to-ripped comparison should also be made before re-ripping your entire video collection. Much like a compressed audio file looses fidelity in the compression process, so does ripping a video file to a compressed format. Compressed output file formats, while much smaller in size, will lose valuable details. For this reason, I recommend using an uncompressed video format knowing that one day a better television will appear on your wall than you now have.

Ripping a 4 Gb Blu-ray disc to a lossless format creates huge files, typically over 40 Gb in size. If you have a large collection of DVDs, Blu-rays, and 4K discs, you need to start adding up the resulting file sizes of these ripped versions when knowing how big of a drive on which you need to store them.

For example, I had about 300 video discs I wanted to keep so simple math showed me that I needed a 2 Tb drive on which to store them (including expansion). Because of these huge requirements, you may choose to add digital video to your library at a later time. Since my video collection was smaller than my audio collection, I chose to put my audio files on one external storage device (an 8 Tb NAS) and my video files on another external storage device (a 4 Tb DAS).

The software that I used is called MakeMKV (http://www.makemkv.com/). As of this writing, the MakeMKV video-ripping software is a beta version and therefore free. I have used this successfully to rip my entire collection with only a few DVD exceptions. The resultant video and audio quality is quite good. As mentioned, other ripping software exists and the same method for choosing the HOME location of your video library should be done for your video ripping software (see Initial Configuration earlier in this chapter).

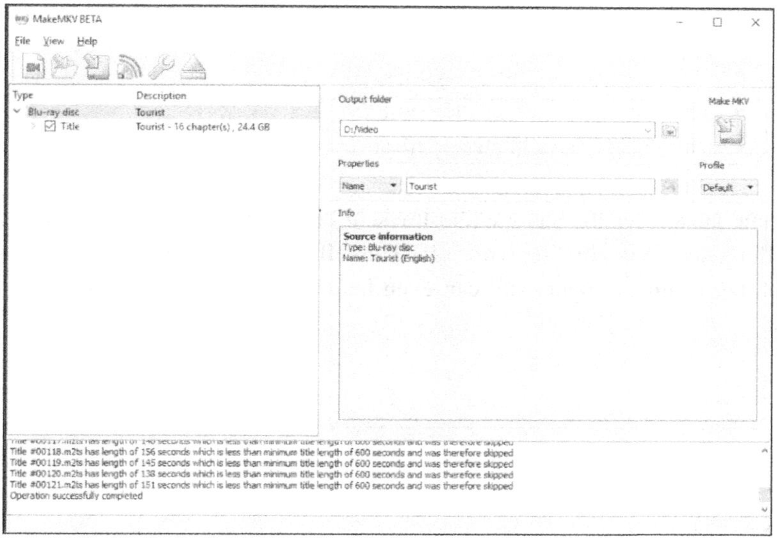

Figure 93 - MakeMKV Video Ripping Software

In the above image, the HOME library is set to D:\Videos and the minimum conversion length of this application is set to 600 seconds (this skips most trailers, clutter, and other commercials as noted in the log at the bottom of the screen).

High-Resolution Audio Download Sites

So far, so good, but the digital audio media you have is not of the highest-quality available. Here is where the high-resolution download services will help you out. But where do you go to get these downloads?

Fortunately, others have done your homework for you and the website. **24bit96.com** maintains a list of high-resolution audio download sites:

https://www.24bit96.com/24bit96khz-download-sites/hd-music-download.html

The following is a short list of my personal favorite sites providing high-quality digital audio media:

- https://bandcamp.com/
- http://bluecoastrecords.com/
- https://burningshed.com/
- http://www.eclassical.com/
- https://www.highdeftapetransfers.com/
- https://www.highresaudio.com/
- https://www.promates.com/music-store/

Some sites offer the same audio track in different formats so you can preview the differences in quality for yourself. This allows you to see if your system can resolve these improvements and can even help you uncover that weak link in your system.

Review

For you, this has probably been a long and sometimes indecisive road where you first thought you understood what it is that you wanted and later found that it was not. Part of the fun of this hobby is making changes and improvements to your system and for some this is a never-ending quest for the search for audio and video perfection.

No matter what type of streaming media system you decided to build, at this point you should have:

1. Assembled the components of your streaming media computer
2. Installed the management software
3. Ripped your existing CD library to a lossless format on your digital storage device, and
4. Ripped your DVD and Blu-ray library to a lossless format on your digital storage device

After ripping your current library to your digital storage device, you may have run out of storage space and needed to re-think your decisions. A DAS allows you to quickly add more storage but is limited by the length of the interconnecting cable. A NAS may be the best solution since some are available with HDMI outputs and built specifically for streaming media.

Planning for growth is also important in anticipating your long-term storage capacity needs. In other words, it is one thing to provide a quick fix to band-aid a problem and another to take time and consider other long-term issues, such as hard drive failures and backups. For this reason, multi-bay DAS or NAS devices offer growth potential where single-bay devices do not.

When selecting the right hard drives for the job you must consider how well they are made. Enterprise-class and Server-class drives offer much better reliability and even self-healing designs than conventional desktop or laptop models.

Setting up your network can start with a good tri-band Wi-Fi router. However, if there are video gamers or multiple concurrent Wi-Fi users in your family, you probably want to move to hard-wired Ethernet connections for your system. CAT-7 and CAT-8 cables offer the best noise immunity and provide the best clarity. You may also consider using a separate (second) LAN for streaming.

Speaking of noise, line conditioners and power conditioners both have their place in assuring the power your system uses is as noise-free as possible. Noise always degrades the audio and video quality and is a good investment.

Another source of noise is found when using multiple outlets to operate all of the components in your system. All outlets should be on the same electrical phase (all on L1 or all on L2). Changing the electrical phase of your outlets requires the services of a licensed electrician but the results are well worth the investment.

Regular backup of whatever storage device you use for your media library are extremely important. Backups can range from simple portable add-on devices (Thunderbolt, Firewire, USB, etc.) or more complex involving multiple NAS devices at multiple locations. Using two identical NAS devices, one at your primary residence and one at your second (or vacation) home, spreads the higher cost of such an elaborate form of redundancy while adding convenience at both locations.

No matter what system you presently have, it can always be improved. Finding where the weakest link hides in your A-V chain is another fun part of this hobby. Adding better quality cables one at a time is one way you can isolate where this weak link exists. Moving to an outboard DAC can instantly transform the sound of your music while offering the ability to decode higher-resolution audio formats.

Migrating from HD to UHD video is still another option that you may incorporate in due time. Storing UHD videos will add huge demands on the storage space in your present system and may require upgrading other devices for best results.

When looking at the weak links in my own system (and truly there are many), I set aside what so-called experts predictably regurgitate and see or hear for myself in an unbiased completely neutral position what changes I see and hear by doing this or adding that. Reading reviews and borrowing equipment is a good way to begin. But even in this approach you can still overlook something based on a personal bias.

Try your best to rise above these biases and embrace your imagination. After all, when removing all limits you open yourself up to unlimited possibilities. The next chapter discusses where to go from here (yes, there is more you can do).

Now What?

If you started this adventure from a hand-held streaming media device, you probably have a different opinion of how this hand-held device serves you. It would be nice to put the exact same quality your home streaming media system offers in your car but that just is not possible. What you can do and what you want to do may not be the same thing.

What you can do is to sit back and appreciate your efforts and think about all of the fun you had when choosing to start on this adventure. Most folks never get up the nerve to do what it is you have just done so for one thing you can consider yourself a part of an elite group. And of course there is the joy you feel when just listening to or watching your efforts come to life.

One thing is certain: the future will change and you have the opportunity to keep up with these changes – or not. When I get my system to a certain point where I am happy with it, I forget about keeping up with improvements and enjoy what I have. After all, that is why I decided to do this in the first place. I recommend that you do the same and take time to "smell the roses."

If you have friends with similar systems, of course you will compare theirs to yours and you may believe that theirs is "better." I encourage you to not pass judgment on your system in such cases but go back and remember why you chose what you did.

There will be times when you have no choice but to replace something like when it breaks down or fails completely. This is the best time to pull out your upgrade plan and see where this particular event falls into your plan. Now you have more fun of deciding what to get that will give you the biggest bang for the buck.

Or you may look at it another way as I often do: Look not at the differences in the components but rather at the differences in the room. I have often said that the room in which your system resides is your system's "missing component." The unknown compromises you have made in your system began early on with the room you chose to watch and listen. Ask yourself this question: *Is there a better location in my home for this system?*

Sometimes you can get things for free. The amount and location of the furniture in this same room influences how your system performs. Moving things around and repositioning your speakers is always a zero-cost change and one that could give you a remarkable bump in its potential.

With few exceptions I have found that when it comes to higher fidelity, less is more. I have seen that taking furniture out of a room, especially on the end with your primary speaker, helps improve the sound in most cases. Adding proper room treatments such as *diffusers* and *absorbers* also will often make huge improvements to the potential of your system's accuracy

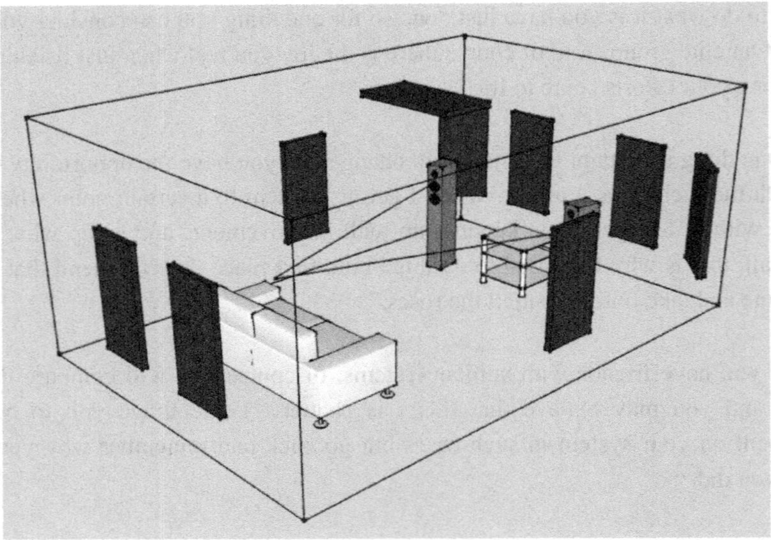

Figure 94 - Acoustic Treatments Can Help

But folding these types of things in an aesthetic manner into your room can be your biggest challenge, especially when you live with another person. Considering their needs and yours is where compromises are struck.

For example, I wanted to put an absorber on the ceiling of my listening room and my wife would not even entertain this thought. To her, it would be too unconventional, visually distracting, and an aesthetic disaster. She was right on all fronts. However, one day I used thumb tacks to temporarily hold in place an absorber to demonstrate the effect I was after. Once she heard the difference (and this difference was striking), we worked together to find a way to make it happen.

No matter what your upgrade plans are, always keep an open mind since the things that are termed "snake oil" today can in fact be proven otherwise tomorrow. As Albert Einstein once said:

We cannot solve our problems with the same thinking we used when we created them.

The moral of this prolonged streaming story is this: be happy with what you have but keep your eyes and ears open for appropriate change. Always consult those who must live with your choices before implementing your plans.

Abbreviations and Glossary

Term	Meaning
AAC	Advanced Audio Coding. A lossy digital audio format (.aac file extension).
Absorber	An acoustic treatment that soaks up sound.
ADC	Analog to Digital Converter
Ai1	All-in-one device. An off-the-shelf unit that allows you to playback digital music and/or digital video files where everything required to do so is in one box.
AIF	See *AIFF*
AIFF	Audio Interchange File Format. An uncompressed digital audio format (.aiff or .aif file extension).
AirPlay	A wireless streaming protocol that permits audio, video, device screens, and photos, together with related metadata between any number of connected devices.
ALAC	Apple Lossless Audio Codec. A lossless digital audio format.
ATRAC	Adaptive Transform Acoustic Coding. A lossless digital audio format with user-defined (lossy) compression rates.
AU	Audio Uncompressed. An uncompressed digital audio format (.au file extension).
Audionirvana	Mac DoP Player.
A-V	Audio-video
Bandwidth	The transfer speed you pay for and receive from your Internet Service Provider. Bandwidth determines how responsive data transfers occur to/from external sources on the Internet to your computer (upload and download speeds).
Bit Depth	The number of digital digits used to record the loudness of an analog signal. Red Book CDs use a bit depth of 16 while others use 20, 24, or more. In other words, larger bit depths record a greater the dynamic range than smaller bit depths. See also *Sampling Rate* and *Resolution.*
Bluetooth	A wireless technology for sending audio from one device to another.
Casting	A wireless technology for sending audio and video from one device to another.

Term	Meaning
CD-A	CD Audio. An uncompressed digital audio format
CODEC	Coder-Decoder. A software technique
Compressed Audio	See *Compression.*
Compression	A method of reducing the size of a stored digital file. Compressed digital files require an expander (or uncompressor) to restore a file to its original content. Compression techniques can be applied to audio and video files.
DAC	Digital to Analog Converter
DAS (*DASD*)	Direct-Access Storage. A digital storage device that directly connects to a computer such as an internal or external drive connected via a USB, Thunderbolt, eSATA, or FireWire cable.
Diffuser	An acoustic treatment that scatters sound.
Digital Word	A combination of two parameters: Sampling Rate plus Bit Depth. The size of a digital word is typically expressed as *Sampling Rate / Bit Depth* such as 44/16 (44 KHz/16-bit) or 192/24 (192 KHz/24-bit).
DoP	DSD over PCM
DSD	Direct-Stream Digital. A lossless digital audio format. It uses multiples (1x, 2x, 4x, or 8x) of a 2.8224 MHz / 1-bit PCM sampling technology and can be abbreviated as such (2.8, 5.6, etc.). DSD labels indicate the sampling rate compared to that of a Red Book CD where DSD64 is really 64x the sampling rate of a Red Book CD or 2.8MHz/1, DSD128 is 128x or 5.6 MHz/1, etc. New developments have increased the bit depth of DSD to 8-bits (wide DSD) thereby improving its resolution and noise.
DXD	Digital eXtreme Definition. An uncompressed digital audio format (.dxd file extension). It uses a 24-bit /352.8 kHz PCM sampling technology.
eSATA	A port used to connect high-speed external digital storage devices to a Windows computer
FireWire	A port used to connect high-speed external digital storage devices to a Windows computer
FLAC	Free Lossless Audio Compression. A lossless digital audio format with user-defined compression rates, sampling frequencies, and bit depths (.flac file extension).
Foobar2000	Windows Audio Player

Term	Meaning
HQPlayer	DoP Player for Mac, Windows, and Linux®
HTPC	Home Theater PC. See *Media Server*
IPTV	Internet Streaming TV.
JRiver	Streaming audio application
Media Server	A computer device (or group of devices) that stores and manages audio and video media. These servers allow you to listen to digital music and watch digital videos via a software application that allows you to hear and see what you have stored in a convenient, user-friendly manner through your other home entertainment devices (receiver and speakers, television, etc.).
MKV	A lossless digital multimedia format (.mkv file extension).
Mirror	See *RAID-0*
MP3	A lossy digital audio format (.mp3 file extension).
MP4	A lossy digital multimedia format (.mp4 file extension).
MQA	Master Quality Authenticated. An encoded form of high-quality Internet streaming audio developed by Meridian Electronics.
NAS	Network Access Storage
Opus	A lossy digital audio format.
OTA	Over-The-Air. Free television station broadcasts in your local area. Stations included are your local major network TV channels such as ABC, NBC, CBS, and Fox plus other smaller independents.
P&P	Plug-and-play
PCM	Pulse Code Modulation. A digital audio format capable of either uncompressed (RAW or header-less) or lossless (header) encoding.
Quantization	A translation of a real value (here an analog signal) to its numerical equivalent (here a digital word). A quantized value is an approximation of this measurement rounded off to the nearest possible value (this approximation is also known as a quantization error). Quantization errors are unrecoverable. In audio, an Analog-to-Digital Converter (ADC) is an example of quantizer.

Term	Meaning
RAID	Redundant Array of Inexpensive Drives. A computer storage technique that employs multiple storage devices (hard disk drives, solid-state drives, or flash drives) to gain more storage space and/or improve reliability. RAID arrays lower the risk of data loss due to the most common type of computer failure: a storage device failure. Also called *RAID Level* when describing the type of array employed. See *also RAID-0, RAID-1*, and *RAID-5*.
RAID-1	(*Mirror*) A storage scheme that employs two devices where the same data is written simultaneously to both devices. Here if one device fails, the array continues to operate from the other device (at a slower rate). The array can be rebuilt by replacing the failed drive without losing data. The maximum capacity of the array is equal to the smallest single device in the array.
RAID-5	(*Striping with Parity*). A storage scheme that employs a minimum of three devices with a software scheme that stores parity information on all three devices (striped). By storing parity information on all devices it allows the array to continue operating (at a slower rate) when one of the three devices fails. The array can be rebuilt by replacing the failed drive without losing data. The maximum capacity of the array is dependent upon the number of devices used in the array (3, 4, 5, etc.).
RAM	Random Access Memory. A high-speed scratchpad area your computer uses to temporarily store data.
Resolution	The ability of a digital word to detect minute changes in the amplitude of an analog signal. Digital Words with greater sampling rates have better resolution than those digital words with lower sampling rates. Digital words with greater bit depths have better resolution than those with lesser bit depths. By definition, an analog signal has an infinite sampling rate and an infinite bit depth. See also *Quantization.*
Sampling Rate	The clocked slice of time at which a sample of an analog signal is taken. Higher sampling rates are more accurate (higher resolution) than lower sampling rates. A digital signal with an infinite number of samples is by definition an analog signal. See also *Bit Depth* and *Resolution.*
SHN	Shorten. A lossless digital audio format (.shn file extension).

Term	Meaning
STB	Set Top Box. A device that enables free and subscription-based streaming video to be viewed on a television. Some STBs also decode low-resolution audio and can stream from DAS and NAS storage devices.
Teac HR Audio Player	Windows DoP Player.
Throughput	A measurement of data speeds within your home network. Throughput determines how fast data transfers occur between any two devices connected to your LAN.
Thunderbolt	A port used to connect high-speed external digital storage devices to a Macintosh computer
TOSLINK (*optical cable*)	*Toshiba Link.* A standardized optical fiber connector system used to transfer digital audio signals. Also called an *optical audio cable.*
TTA	TrueAudio. A lossless digital audio format (.tta file extension).
Vorbis	A lossy digital audio format.
WAV	Waveform Audio. An uncompressed digital audio format (.wav file extension).
WavPack	A lossless digital audio format.
Wi-Fi	A wireless technology for sending digital information from one device to another through an intermediate device.
WMA	Windows Media Audio. A lossless digital audio format (.wma file extension).

About the Author

Philip Rastocny started his musical appreciation career by learning how to play the accordion. Coming from a Czechoslovakian family, it seemed to his parents that this instrument would be the right choice for weddings and parties of which they frequented. After all, with a child able to perform polkas, what else does a family need to dance and have fun at a Slovak celebration?

In his senior year in high school, he ventured down to a teacher's office where he was introduced to his first audiophile system: a simple tube receiver coupled to an analog turntable and tiny bookshelf speakers. Although modest in cost, it produced better sounding music than anything he had heard to date.

He never forgot that unique difference and immediately got hooked into the high-end. While in the military, he allocated a portion of his income to improving his system and soon a similar version of these high-school inspired components appeared in his room. After marrying, his wife once told a friend that "I married him for his stereo…" although she adds today that there was a little more to it than just that.

After the military and while in college, he secured a job in the audio department of a PX on a local Air Force base and a friend loaned him a copy of the magazine called *The Absolute Sound*®. Reading about what people heard rather than quoting specifications was a new thing to him and *finally* reviewers were talking about the things he appreciated about music reproduction. He immediately subscribed to this magazine.

In college he studied Electrical Engineering and Bio-Medical Engineering and graduated with two degrees. Armed with the knowledge of the internal workings of audio, he quickly dismantled each piece of his gear tweaking this and modifying that. He built a pair of speakers thinking "How hard can it be to stuff speakers in a box?" and quickly learned why prices were so high; it was simple to make sound but far more difficult to get things to sound real.

After 34 versions of one speaker, he became satisfied with it and turned his attention to the recording chain. At the time direct-to-disc recordings were the best signal source available but there was still a marked difference between live performances and recorded music. He then designed and built a portable recording studio and started making his own live recordings. Listening to these proved to him that the problem was not in the playback systems but in the recording end.

But setting up a session at the University of Denver or some local church for an organ recital took *a lot* of time and effort and although the results were astonishing the cost was more than the relationship with his wife could bear. One day while tweaking the network in his system she came down and emphatically proclaimed one word "Enough!" and that was that. He stopped cold turkey and appreciated for decades what he had accomplished. Many years of listening pleasure ensued but the bug stayed with him; music sounded good but still not real.

He then sold his home and down-sized into an RV and the challenge of good music became compromised. Although there was room in his big rig to add good equipment, the quality of 12-volt components was abysmal. Again after a few months of redesigning and tweaking he emerged with an integrated audio-video system that was indeed amazing for its size. But life on the road was short lived when his wife's sister contracted cancer. They decided to drop their easy life as wandering gypsies and move to Nashville to support her sister's recovery.

This was a good thing since Nashville is the Country Music capitol of the world. Exploring the local symphony also bore fruit; this is where he discovered Maestro Schermerhorn who turned out to be one of his idols.

With his sister-in-law slowly recovering, they bought a modest home in one of the bedroom communities of north Nashville where they parked their RV and reacquired their love for a home theater. Pulling their old gear out of storage, more tweaking came into play and the turntable took a back seat while a CD carousel stored discs. Two years later, they hit the road again and landed in central Florida.

This is where streaming digital music found its way into their home theater. After joining a local audio club and being exposed to all sorts of high-end streaming gear, the thought crossed his mind that there may be a benefit into folding such a feature into his analog-only system. But the bar was set pretty high with the evolution of his hand-made equipment. With the encouragement of his wife, he slowly integrated one approach and then another and another until he finally understanding where the real issues hid in digital reproduction.

In the process, he dispelled many myths and misunderstandings of where these issues arose and hopes that this guide gives you the same peace he now has.

Made in the USA
Monee, IL
04 November 2026

67387008R00075